FOR IAN AND JENNY

THANKS FOR EVERYTHING!

CONTENTS

PREFACE TO THE ROUTLEDGE CLASSICS EDITION

As its title shows, this book deals in the first place with the alleged clash between 'the two cultures' (humanities and science). It ends up, however, by ranging a good deal wider and considering the general use of the imagination in human life.

Although the two-cultures clash had bothered me for a long time it was only during the 1990s that two particular salvoes in it finally roused me to try to do something positive about it. One salvo came from the chemist Peter Atkins, who proposed roundly that poetry was quite useless since it could never add anything to the omnicompetent revelations of science. The other came from Richard Dawkins, who was much more sympathetic to literature. Indeed he seemed actually fond of it. But he was puzzled and distressed to notice how hostile certain poets such as Keats, Blake and Wordsworth had apparently been to certain scientific points of view. Why (he asked) could they not have been more enlightened? How could they be so perverse?

In Part 1 of this book I discuss these two manifestoes and try to explain the long history that lies behind them. The manifestoes themselves struck me as deserving attention because the attitude behind them is one that a lot of other people probably share today, though they may not express it so plainly. Its central idea is, I think, a conviction that there must be a single human faculty that is finally able to solve all our problems and arbitrate all our disputes. And today the intellect, as exercised in science, is most often thought to be that faculty. It seemed to follow, then, that other human powers and pursuits ought not to be allowed to compete with it and that we probably should not waste too much time in exercising them. Science should simply be acknowledged as a benign and absolute ruler.

I do not myself think that this monarchical vision is very realistic, even as such imaginative visions go. But the first point that I wanted to make about it was that it is indeed a *vision* – a general way of looking at the world, not itself a piece of scientific reasoning. It is one possible imaginative structure among other available ones that we can use. And this particular structure, as we have it today, is certainly not something that has emerged from science itself. It actually derives from the other side of the gulf. It has been the creation of certain past philosophers and poets – of the Greek Atomists, of Lucretius, Bacon, Locke, Laplace, Comte and other enlightenment sages, ably assisted by many Augustan and contemporary writers, down to Bertrand Russell, Karl Popper and Jacques Monod.

In short, like any other background picture that we use in our efforts to make sense of things, this vision is an imaginative product with a rich social background. It is a pattern that has recommended itself, for various reasons, to powerful tides of feeling in the course of history. Its articulate form, however, derives from particular past disputes among philosophers who expressed that feeling. If, then, we really want to understand it – if we want to grasp its meaning rather than just repeating a message passed on to us by earlier thinkers – we must investigate what those philosophers were up to. The alternative is to remain in bondage to their thinking. And the disputes involved here have themselves been highly diverse – so diverse that they can easily suggest the programme of a football League. There have been battles of Science v. Religion, Reason v. Feeling, Classical v. Romantic, Masculine v. Feminine, Materialist v. Idealist, Body v. Mind.

All these disputes, various though they are, share one unfortunate feature: they all tend to drive us into inner conflict. They all ask us to take sides between two aspects of our inner selves rather than finding a way to reconcile them. They all ask us to envisage ourselves – our own identity – in a particular way that will suit their own side of the argument. They set body against mind, feelings against reason, male against female.

These inner conflicts tear our lives apart and it is surely most important to resist them. In Part II, accordingly, I have tried to show the bearing of these disputes on our sense of personal identity – on what Jung called the integration of the personality. I have tried to show ways in which we can resist contemporary ideas that seem to split us in two, ways in which we can combine the best insights of the two contenders. And because academic specialisation tends today to drive still further wedges between these conflicting demands, offering different, incompatible images of our condition, I have suggested that we need to resist this fragmentation strongly:

> There is nothing wrong with such images, but . . . no one of them can ever serve for all purposes. No picture should be allowed to become an imaginative monoculture.

We need always to be looking for a deeper unity.

In Part III, therefore, I go on to consider a further crucial aspect of that unity – the unity between ourselves and the living world around us.

One very deplorable feature of the simple, one-sided, reductive attitude that we have inherited from the seventeenth century has been a detached, contemptuous attitude to the natural world – a vision of ourselves as something quite distinct from that world, superior to it, able to study and use it just as we please. As we are now beginning to realise, this is a point on which Enlightenment thinking has been terribly faulty. To correct this bad imagery we need (again) not just arguments but a better, more realistic image. We need, in fact, an image that shows us justly as what we are – a miniscule, wholly dependent part, organically situated within a far greater whole, a biosphere that we can easily damage, but which we could never direct or replace. And for this purpose (as I have suggested) we will do well to use the concept of Gaia.

ACKNOWLEDGEMENTS

Part 1 of this book has grown out of two Victor Cook lectures (called 'Science and Poetry' and 'Atoms, Memes and Individuals') which the University of St Andrews kindly invited me to give in 1998. The generous terms of this foundation allowed me to give these lectures in Leeds and Glasgow as well as at St Andrews itself, a friendly arrangement which gave me the benefit of much helpful discussion in all three places. I am grateful to John Haldane and to everybody else, both at St Andrews and at the other two universities, who made these occasions so lively and fruitful. It was at this stage that I began to see how several (apparently separate) lines on which I had been working might converge to form this book.

In Part 2, Chapters 7 and 8 have grown out of a shorter version, 'Putting Our Selves Together Again' which appeared in *Consciousness and Human Identity* (edited by John Cornwell, Oxford University Press, 1998). Chapters 10 and 11 are an expanded version of 'Consciousness, Fatalism and Science' which appeared in *The Human Person in Science and Theology* (Edinburgh, T&T Clark,

2000). Chapters 12 to 16 incorporate two articles published in the *Journal of Consciousness Studies* – 'One World but a Big One' in vol. 3, no. 5–6, 1996 and 'Being Scientific about Our Selves' in vol. 6, April 1999.

In Part 3, Chapters 17 and 18 enlarge on an article called 'Towards an Ethic of Global Responsibility' which appeared in *Human Rights in Global Politics* (edited by Tim Dunne and Nicholas Wheeler, Cambridge University Press, 1999). The remaining five chapters are based on a pamphlet brought out by DEMOS in May 2001 and called 'Gaia: The Next Big Idea'. I am grateful to all the publishers and editors involved for making possible these earlier publications and for allowing versions of them to be reprinted here.

I have, of course, had endless help from friends, colleagues and fellow-explorers of this vast territory, notably from John Ziman, Steven Rose, Brian Goodwin, James Lovelock, Raymond Tallis, Rom Harré, Mary Clark, Anne Primavesi and Andrew Brown. Claire Lamont has been most helpful in saving me from major errors about current literary criticism. Strachan Donnelley and his colleagues at the Hastings Center have repeatedly reminded me of ways in which the world is even wilder than I had been supposing. And my philosophical colleagues in Newcastle who have survived the destruction of their department – Ian Ground, Judith Hughes, Willie Charlton and Michael Bavidge – have continually thrown light on my awkward problems. So have David Midgley and David Brazier, who have also – on top of contributing a useful Buddhist angle – given me invaluable help, along with Tom and Martin Midgley, in baffling the malice of the Word Processor.

I have to thank the Society of Authors, which is the literary representative of the Estate of A.E. Housman, for permission to print a poem from *A Shropshire Lad* which appears on p. 124.

INTRODUCTION

WHO ARE WE?

This is a book about personal identity, about who and what we are. It is about the unity of our lives. It tries to suggest how we can resist the academic fashions that now fragment us. It looks at the various aspects of our selves which get separated by being involved in the different arts and sciences, and it asks how we can bring them together within our wider life. In particular, it asks how we can bring together our ideas of *science* and *poetry* within a whole that has a place for both of them. It investigates the strange, imperialistic, isolating ideology about science which now makes this kind of connection seem impossible.

That ideology is what makes science itself look alarming today. What is called 'anti-science' feeling is not usually an objection to the actual discovery of facts about the world. (That would be very odd.) Instead, it is a protest against this imperialism – a revulsion against the way of thinking which deliberately extends the impersonal, reductive, atomistic methods that are appropriate to physical science into social and psychological enquiries where they work badly. That they do work badly there has often been pointed out. Yet these methods are still often promoted as being the only rational way to understand such topics.

This deliberate extension makes it seem as if something called *science* is forbidding us to be human. But science does no such thing. The call to extend its methods into unsuitable territory does not come from science itself but from a peculiar vision of the world, a set of imaginative habits that have been associated with modern science since its dawn in the seventeenth century. Our visions – our ways of imagining the world – determine the direction of our thoughts, as well as being the source of our poetry. Poetry exists to express those visions directly, in concentrated form. But they are also expressed less directly in all our thoughts and actions, including scientific ones, where they often pass unnoticed and uncriticised.

SCIENTIFIC PSYCHOLOGY?

I have suggested that the particular vision which has been seen as scientific here centres on an unbalanced fascination with the imagery of atomism – a notion that the only way to understand anything is to break it into its ultimate smallest parts and to conceive these as making up something comparable to a machine. Because that method succeeded for a time so well in the physical sciences, people have hoped to extend it to the rest of life in two ways. The first and more obvious way is by reducing mind itself to matter and thus to physical particles. This is seen as a way to mend the yawning division which Descartes introduced between mind and body by letting the major partner swallow up the minor one. Thus, psychiatrists have sometimes tried to view their patients merely as physical mechanisms and behaviourist psychologists hoped to study human life purely in terms of outward behaviour – of the movement of human bodies – without referring at all to the thoughts and feelings of the people involved.

Not surprisingly, that project worked badly and the behaviourist scheme has been abandoned. Yet the underlying dream of making psychology *scientific* in some related sense still persists. That dream is not centrally fuelled by the wish for knowledge. It is primarily a dream of taming and simplifying our inner life so that it will somehow conform to the known laws of matter and will stop setting us problems of its own. The sciences chiefly invoked at present for this project are neurology and the study of evolution. This means that the best way to study people is by looking at them either through a microscope or through the wrong end of a telescope – the telescope of evolutionary time.

These are indeed both useful methods. But it is a little odd to give them this kind of priority. The reason for preferring these studies seems to be much more because of their success in other fields than from any special likelihood that they will help us here. This is essentially the approach well described of late by the story of the man who is found looking for his keys under a street-lamp and is asked whether that is where he dropped them. 'No,' he says, 'but it's much the easiest place to look.'

The other way of atomising human life does not involve materialism. It is *social atomism* – individualism – the idea that only individuals are real while the groupings in which they live are not. Each citizen is then a distinct, ultimately independent unit, linked to the others around it only externally, by contract. The roots of this idea are of course political rather than scientific. But individualist theorists have for some time claimed that the view is scientific in a sense that roots it in physical science. Nineteenth-century social Darwinists did this by insisting that free commercial competition was the predestined spearhead of the whole evolutionary process. In our own day, the rhetoric of the

Selfish Gene has a similar effect, though it is less explicitly political.

REDUCTION, REALITY AND POWER

Both these atomistic doctrines rest on the idea that competition between separate units is the ultimate law of life. Both ignore the obviously equal importance of co-operation between organisms – and between the parts of organisms – at all levels. Both confer a misleading air of scientific rigour on the proposition that there is, ultimately, no such thing as society. Both therefore depend on a rather odd piece of metaphysics, namely the 'reductive' assumption that certain parts are, in some sense, always more real and significant than the whole they belong to. Thus Richard Dawkins:

> The individual organism . . . is not fundamental to life, but something that emerges when genes, which at the beginning of evolution were separate, warring entities, gang together in co-operative groups, as 'selfish co-operators'. The individual organism is *not exactly an illusion*. It is too concrete for that. But it is a secondary, derived phenomenon, cobbled together as a consequence of the actions of fundamentally separate, even warring agents. I shan't develop the idea but just float . . . the idea of a comparison with memes. Perhaps the subjective 'I', the person that I feel myself to be, is the same kind of semi-illusion . . . *The subjective feeling of 'somebody in there' may be a cobbled, emergent, semi-illusion* analogous to the individual body emerging in evolution from the uneasy co-operation of genes.
>
> (From *Unweaving the Rainbow*, London, Penguin, 1998, pp. 308–9; emphases mine)

This is the formula of metaphysical reduction and it needs

explanation. What can it actually mean to suggest that the things that we directly deal with are in some sense *less real* than certain selected parts – or alleged parts – of them? This mysterious point is seldom spelt out but it appears to centre on causality. The suggestion is that only these special parts are causally active. They are spontaneous, self-moving movers, while the wholes that they compose are mere passive outcomes of their activity.

Dawkins' wording here suggests that this is a historical truth – that these parts actually existed on their own before these wholes and gave rise to them. But this is not literal fact; it is a piece of symbolism. Memes, if they can be said to exist at all, certainly do so only as emergent aspects of human social life. Even their most fervent supporters have not suggested that they pre-existed as spiritual beings who originally produced that life.

We will come back to these ambitious memes in Chapter 5. I have discussed them more fully elsewhere.[1] It is also worth noticing in this connection the oddity of current doctrines about Evolutionary Psychology, which atomise the mind itself into a string of separate molecules, a module for each capacity, envisaging the whole group of capacities as comparable to a Swiss Army Knife. But a Swiss Army Knife is a tool. The Swiss Army, in fact, does not consist only of knives. It also needs people who know how to use them – people who can choose between the different blades. And the business of psychology (as is usually thought) is to understand people, not their tools.

As for genes, it is not in fact seriously suggested that, as a matter of historical fact, they ever existed as independent items, precursors and architects of the organisms that now embody them. As is well known, DNA itself is a totally inert molecule which would never have done anything if it had been put down in a world without organisms. It is produced by living cells just as their other essential molecules are and it works only as a part of them. It is no more capable of going around on its own than bones or leaves are. As Steven Rose explains:

What brings DNA to life, what gives it meaning is the cellular environment in which it is embedded ... Genetic theorists with little biochemical understanding have been profoundly misled by the metaphors that Crick provided in describing DNA (and RNA) as 'self-replicating' molecules or replicators, as if they could do it all by themselves. But they aren't and they can't ... You may leave DNA or RNA for as long as you like in a test-tube and they will remain inert: they certainly won't make copies of themselves ... The functioning cell, as a unit, constrains the properties of its individual components. The whole has primacy over its parts.

(Steven Rose, *Lifelines: Biology, Freedom, Determinism*, London, Penguin, 1997, pp. 127 and 169)

This well-known fact has been obscured in recent times by the enormous interest that was naturally, and rightly, generated by the discovery of DNA's role in reproduction. A more profound cause of it, however, is the symbolism – the uncontrolled tangle of metaphors that has grown up around that discovery, building up the notion of genes as possessed of an almost magical, spontaneous power. This exciting idea has been made even more seductive by the hope that, through genetic engineering, we ourselves may be able to dominate these powerful genes and thus become controllers of the whole system.

This misleading sense of genes as all-powerful has been much encouraged by the information-metaphor which depicts them as constantly giving orders to the entities around them. Rose writes:

To continue the linguistic, information-theory metaphor within which genetic theory was now to be formulated, the directed synthesis of RNA on DNA was termed *transcription*, and the synthesis of protein on the RNA was *translation*. DNA had become the master-molecule, and the nucleus in which it was located had assumed its patriarchal role in relation to the rest of the cell. It is hard to know which had more impact on the future directions of biology – the determination of the role of DNA in protein synthesis, or the organizing power of the metaphor within which it was framed.

(ibid., p. 120)

These linguistic images, when taken seriously, quickly made it seem that, as Dawkins puts it, 'Life is just bytes and bytes and bytes of digital information'.[2] That is, it is just a long string of orders through which the gene – always the informant – tells the docile cells what to do. But that, as Rose points out, is not at all how living cells actually work:

Far from being isolated in the cell nucleus, magisterially issuing orders by which the rest of the cell is commanded, genes … are in constant dynamic exchange with their cellular environment. The gene as a unit determinant of a character remains a convenient Mendelian abstraction, suitable for armchair theorists and computer-modellers with digital mind-sets. The gene as an active participant in the cellular orchestra in any individual's lifeline is a very different proposition … The organism is both the weaver and the pattern it weaves, the choreographer and the dance that is danced.

(ibid., pp. 125–6 and 171)

DIFFERENT LEVELS, DIFFERENT PATTERNS

Why, then, have biologists lately become so obsessed with genes? Of course their interest has had plenty of point because the discoveries connected with genetics have been of real importance. But, like many such swings of interest in science, this one has led to a crescendo of over-simplification and exaggeration. As Brian Goodwin puts it:

> A striking paradox that has emerged from Darwin's way of approaching biological questions is that organisms, which he took to be primary examples of living nature, have faded away to the point where they no longer exist as fundamental and irreplaceable units of life ... Modern biology has come to occupy an extreme position in the spectrum of the sciences, dominated by historical explanations in terms of the evolutionary adventures of genes. Physics, on the other hand has developed explanations of different levels of reality, microscopic and macroscopic, in terms of theories appropriate to these levels, such as quantum mechanics for the behaviour of microscopic particles ... and hydrodynamics for the behaviour of macroscopic liquids.
>
> (*How the Leopard Changed its Spots*, London, Weidenfeld and Nicolson, 1994, pp. ix–x)

Physicists, in fact, no longer make this crude reductionist move of claiming that their latest-discovered entities are the only real ones, because physics – far ahead of biology – has already gone through the trauma of understanding that there are many such discoveries and many such entities. 'Reality' turns out to contain many different kinds of pattern at different levels. No one of these discoveries therefore should be expressed in the dramatic metaphysical language of reality and illusion. Different ways of thinking co-exist and are appropriate on different scales. No one of them dominates or invalidates the others. Accordingly, Goodwin goes on:

> Despite the power of molecular genetics to reveal the hereditary essences of organisms, the large-scale aspects of evolution remain unexplained, including the origin of species ... It is here that new theories, themselves recently emerged within mathematics and physics, offer significant insights into the origins of biological order and form. Whereas physicists have

traditionally dealt with 'simple' systems in the sense that they are made of few types of component, and observed macroscopic or large-scale order is then explained in terms of uniform interactions between these components, biologists deal with systems (cells, organisms) that are hideously complex . . . However, what is being recognised within these 'sciences of complexity', as studies of these highly diverse systems are called, is that there *are* characteristic types of order that emerge from the interactions of many different components . . . Order emerges out of chaos.

(ibid., pp. x–xi)

In short, reductionism is not the only rational way of dealing with differences of scale. There are much better ways of representing them. Different forms of order can co-exist at different levels, so scientists can use different ways of thinking about them without fighting, without insisting on reduction and without scandal. Goodwin comments:

Conflict only arises when there is confusion about what constitutes biological 'reality'. I take the position that organisms are as real, as fundamental, as irreducible as the molecules out of which they are made. They are a separate and distinct level of emergent biological order, and the one to which we most immediately relate since we ourselves are organisms.

(p. xii)

There is, in fact, no need to talk about reality here at all.

I cannot here go further into the fascinating topic of how life actually did originate. (Interested readers should look at Lynn Margulis's discussion of this.)[3] What we need to notice here is how hard it is to fit together the various kinds of atomism that have been introduced into our thought at different organisational levels. If organisms are semi-illusory in relation to genes, are genes (then) also semi-illusory in relation to atoms and quarks? Is nothing actually real but quarks – or whatever particles may succeed quarks after the next revolution in physics? What would that mean? The problem is even worse about the question of society. Social atomism views individual people as autonomous ultimate units in full charge of their destiny. Physical atomism, by contrast, dissolves these people away into chance collections of smaller units such as molecules, quarks or genes, collections that are continuous with the landscape around them. It sees them as subordinate cogwheels exercising no sort of individual control. For the first story, free will is essential. For the second, it is impossible.

Yet both these opposite models of our selves are equally powerful in the rhetoric of today. We are continually being called, on the one hand, to exaggerate our freedom boastfully – which leads to orgies of remorse – and, on the other, to admit that we are actually only helpless cogs. These two exaggerations have, of course, grown up in reaction against one another. When we oscillate helplessly between them we manage to get the worst of both worlds.

BEYOND ATOMISM

This book is an attempt to find ways of avoiding that fate. It tries to understand better the general way in which these imaginative visions work and, more particularly, to grasp the part which atomistic visions have played in shaping our own culture. Part 1 of the book centres on this theme. It shows how, in spite of the clash just noted between them, the two forms of atomism have been closely linked in our history. They have constantly strengthened one another because the surface likeness between their forms has been much more noticed than their incongruity. Thus, the social development of individualism increased the symbolic appeal of physical atomism, while the practical successes of physical atomism made social individualism look scientific.

The social extension of atomistic methods (to which we will return in Chapter 15) is not, of course, really a scientific project at all, though it uses scientific language. It is a distortion that tends to discredit the whole idea of science by exploiting it to draw dubious political and moral conclusions. This distortion itself has become obvious over the very notion of an *atom* – the idea of an impenetrable, essentially separate unit as the ultimate form of matter. We know that today's physicists no longer use this billiard-ball model. They now conceive of particles in terms of their powers and their interactions with other particles, not as inert separate objects. The seventeenth-century idea of a world constructed out of ultimately disconnected units has proved to be simply a mistake. Instead, physicists now see many levels of complexity, many different patterns of connection.

At an obvious level it follows that we ought no longer to be impressed by social atomism, or by behaviourism, in the way that we once were. We can see now that it cannot have been scientific to impose on social affairs a pattern which turns out to have been so inadequate for physics. But the moral goes much deeper. It is one that would still hold even if physics had not changed. That moral is that, quite generally, social and psychological problems cannot be solved by imposing on them irrelevant patterns imported from the physical sciences, merely because they are seductively simple.

Of course simplicity is one aim of explanation. Of course we need parsimony. But it is no use being parsimonious unless you are relevant. Explanations must be complex enough to do the particular work that they are there for, to answer the questions that are actually arising. There are always many alternative ways of simplifying things and we have to choose between them. The kind of parsimony that is too mean to deal with the points that really need explaining is not economy but futile miserliness. For any particular problem, we need a solution that sorts out the particular complications that puzzle us, not one that ignores them because they are untidy.

BECOMING CONSCIOUS OF CONSCIOUSNESS

In the last few decades, one complication of social life that had long been carefully ignored has managed to escape and erupt onto the academic scene. The modest fact that we are conscious is now agreed to constitute 'the problem of consciousness'. It is not really a single problem but an aspect of a thousand problems – namely, their subjective aspect. That aspect was long concealed and suppressed because it was believed that it would be disgracefully subjective even to mention anything subjective – that, in fact, it was impossible to think objectively about subjectivity. (This is the same reasoning which Dr Johnson neatly parodied in the line 'Who drives fat oxen should himself be fat'.)

This veto has now been withdrawn. Scientifically minded people now admit that conscious subjects exist and may affect the world. The first-person point of view is, then, not a myth (as the behaviourists sometimes said) but is, however regrettably, a natural fact like any other and perhaps an important one. The problem then arises: how can we fit it into conceptual schemes that were never meant to accommodate it? How are we to talk about ourselves as subjects? How, in particular, should we talk about the relation between ourselves as subjects and as objects – between the first- and third-person aspects of ourselves? What sort of beings do we – as a whole – now turn out to be? The second part of the book deals mainly with this problem.

There is a real difficulty here because the natural sciences are wholly dedicated to talking about objects. That is their job. People like Galileo laid down clear conventions at the dawn of modern science to exclude everything subjective from those sciences. They cannot, therefore, provide a language for discussing the relations between subjects and objects. This does not, of course, stop scientists discussing these matters. They can perfectly well do so. But in order to do it they, like everyone else, have to use terms drawn from contexts other than the natural sciences, often ones drawn from everyday life.

Many people, however, are convinced that rational, intellectually respectable discussion can only be carried on in scientific language, meaning by *scientific* not just *disciplined* and *methodical*, like the language of history or logic or linguistics – which would be uncontroversial – but *drawn from the natural sciences*. They are sure that – as Richard Dawkins has recently put it – 'Science is the only way we know to understand the real world.'[4] They therefore see the problem of consciousness as essentially one of devising a 'science of consciousness', one which will be either directly derived from the existing natural sciences or else so like those sciences formally as to take its place among them without causing a scandal. Thus, the University of Arizona, when it kindly invited us all to its prestigious Fourth Tucson Conference on Consciousness for the year 2000, began its notice thus:

> Recent years have seen an explosion of work in the sciences and humanities on *science's last great frontier, the problem of consciousness*. Can there be *a scientific theory of consciousness*? If so, what form should this theory take? . . . A special focus of the conference will be the question of how the first-person and third-person perspectives can be integrated, and on *how first-person data on consciousness can be rigorously incorporated into science*.
>
> (Emphases mine)

As we know from discussions of space-travel, this metaphorical talk of 'last frontiers' is always imperialistic. It signals an intention to conquer the outstanding area and bring it under control. These organisers, then – even though they acknowledge contributions from the humanities – still seem to be using a simple territorial map on which any decent conceptual scheme will have to be one 'rigorously incorporated into science'. It is a map which shows science as isolated in purdah, a country cut off, as by an iron curtain, from the rest of our intellectual life. Throughout this book, and especially in Chapter 14, we will be seeing how misleading this is. That restriction prevents us both from appreciating the real importance of science itself and from approaching the large problems about ourselves that we now face.

Investigators using this map approach their new problem on the jigsaw principle, armed with puzzle-pieces from various existing physical sciences such as neurology, quantum mechanics, genetics or the study of evolution. (Rather surprisingly, computer science too is now allowed to count as a science for this purpose, though it has no physical subject-matter, being actually a species of applied logic.) They try to fit their chosen pieces into the problem. But the problem does not accommodate them because it is one of a quite different kind. It is actually about how to relate different puzzles. It concerns *how best to fit together the different aspects of ourselves – notably, ourselves as subjects and ourselves as objects, our inner and our outer lives.*

MINDS AND BODIES: NEITHER APARTHEID NOR CONQUEST WILL WORK

Descartes notoriously simplified this issue in the early days of modern science by sharply dividing mind from body, subject from object, and handing the body over to physical science. This apartheid was a convenient arrangement for many purposes, allowing the different kinds of study to develop separately. But the lack of any intelligible relation between them made it impossible to fit them together. Yet such fitting was needed because, as both kinds of study developed, clashes arose over all sorts of issues where mind and body appeared to interact, centrally over free will. To arbitrate them, the two provinces had somehow to be related.

How can we now deal with these clashes? Descartes' model always had the drawback of suggesting that they could somehow be resolved by conquest – by one partner's swallowing up the other. Mind and body were both called *substances*. Though they were supposed to be totally unlike, this seemed to suggest that they were somehow comparable stuffs, one of which would turn out to be a form of the other. Either matter was really constructed out of mind (idealism) or mind was constructed out of matter (materialism).

Today a vague impression exists that materialism has won this battle. But I think it has become clear that both these solutions are equally unworkable. We have to avoid dividing ourselves up as Descartes did in the first place. *Things go wrong as soon as we start thinking about mind and body as if they were both objects* – that is, separate things in the world. The words *mind* and *body* do not name two separate kinds of stuff, nor two forms of a single stuff. The word *mind* is there to indicate something quite different – namely, ourselves as subjects, beings who *mind* about things. The two words name points of view – the inner and the outer. And these are aspects of the whole person, who is the unit mainly to be considered.

Words like mind and body do not have to be the names of separate items. They, and the other many-sided words that we use for these topics – words such as *care, heart, spirit, sense* – are tools designed for particular kinds of work in the give-and-take of social life. They are essentially vernacular, and that is just their strength. They have been shaped by the everyday context of experience, which is just what we are trying to talk about. When we use them in controversy it is no use trying to disinfect them by the kind of abstraction that Descartes used or by replacing them with invented terms. They not are a cheap substitute, an inadequate 'folk-psychology', due to be replaced by the proper terms of the learned. They are well-adapted tools, shaped in each culture through long experience to express human thoughts and feelings. They come out of the same underlying world-visions which also emerge both in poetry and in science. Any faults that such words have are the faults of those wider visions and of the ways of life that go with them, not symptoms that the words themselves are too crude.

MATERIALISM IN DIFFICULTIES

Among these influential world-visions, the atomistic and mechanistic one that I just mentioned still seems, when seen from a distance, to hold the same prominent and respected place in our culture today that it has occupied since the seventeenth century. But if we look more closely we can see that it is in deep trouble. It is a vision of which far too much has been expected. As always happens in such cases, it started to reveal serious faults at the point when its supporters stopped treating it as just one interesting and fertile suggestion among others and decided to enthrone it as an 'omnicompetent' universal method. Under that dangerous spotlight its various parts began to clash visibly both with one another and with other recognised truths. Confusion is now so bad that an overhaul is unavoidable.

The trouble is particularly serious over the concept of *materialism*. Starting from Descartes' division, this word seems to mean that we should no longer believe in two substances but only in one, namely matter. But in that case, who is there to do the believing? Matter is strictly defined in this system as mere object, passive and inert, not the kind of thing that could possibly think, feel or believe. And the world of matter is supposed to constitute an entirely self-contained machine. Why, then, does all experience show that we ourselves often do think and that our thought affects our actions?

This is the difficulty that has now forced theorists to attend at last to the problem of consciousness. They would probably have done so earlier if they had not got side-tracked by arguments about religion. Many people welcomed materialism primarily as a way to get rid of religion, a reason for disbelieving in God and the immortality of the soul. Exciting political battles could be waged with the churches about this, so such debates held a huge attraction. But, like many lively feuds, these debates have actually been a side-show, a displacement activity, a distraction to avoid the real difficulty. Souls do not only concern us after death. They concern us now because we are conscious now. This fact has to be fitted into the world where we already live. No degree of scepticism about other possible spiritual worlds makes any difference to it.

Traditional materialism, in fact, asks us to believe in a world of objects without subjects, and – since we ourselves are subjects, being asked to do the believing – that proposal makes no sense. This vision is no more plausible than the idealist alternative of subjects without objects, indeed it is actually less so. The trouble is quite simply that the Cartesian concept of *matter*, which was framed in the first place as a contrast to mind, cannot be extended to take in its opposite without losing its meaning. In order to be stretched in that way, it would need to be entirely reshaped. As it happens, theoretical physicists are actually now engaged in reshaping that concept for a number of reasons, two of which are sharply relevant to this topic. One of these is their rejection of traditional determinism. The other is a difficulty about the status of 'observers' who are apparently subjects. Physicists, in fact, now find the seventeenth-century vision of matter unusable for their current purposes and they want to devise a new conceptual scheme to replace it. As we shall see, in attempting this they tend now to reject terms such as materialism altogether.

Many biologists and social scientists, however, do not seem to have yet heard news of this change in physics. They still vigorously promote traditional mechanism, atomism and materialism, along with the determinism that went with them. They still view these as 'hard' and clear doctrines, indispensable elements of rationality. Difficulties have long been obvious about this position. But on top of these familiar troubles, such defenders now face the new obstacle raised by debates about the status of subjectivity itself. Homo sapiens scientificus has started to admit that he is himself conscious and even that some other animals may be so too. This presents traditional-minded biologists and social scientists with serious problems. They are trying to confront them in a number of remarkable ways that will concern us throughout this book.

The obvious central difficulty concerns free will and the reality of human action. When Brutus murdered Caesar, did his own conscious thoughts and feelings contribute nothing to his action? Were those thoughts and feelings (as is now often claimed) merely an extra, a side-effect, a futile spin-off from autonomous processes in the brain? Were the neurones (in fact) the only real actors? If it is held that they were, then this supposedly rational set of doctrines (materialism, atomism, mechanism and determinism) brings with it a much less rational-looking companion, namely fatalism. In this case it doesn't matter at all what we think and feel, because our thoughts and feelings cannot have any consequences. Effort – which is essentially conscious – is ineffectual and we can stop bothering with it. From now on, our conscious selves can just sit back and let the neurones live our lives for us.

WHAT IS FREE WILL FREE FROM?

We will discuss this problem in Chapters 9–11. But it may be as well to say something at once here about how I shall approach it. People discussing free will often take it for granted that determinism is a clear and rational doctrine while the idea of freedom is cloudy and dubious. But actually both ideas, in the form in which they are now usually contrasted, are at least equally obscure. I sign up here with Peter Strawson, who opened his very helpful remarks on the topic by saying 'I belong to . . . the party of those who do not know what determinism is'.[5] Centrally, the trouble is that the word *determine* is so ambiguous. The sense in which a general 'determines the fate of a private' describes outside compulsion. The sense in which 'three points determine a plane' does not involve it. It is this outside compulsion that we need to be free from. The general is a different person from the private. But body and mind are not separate persons. So it is not obvious that any such compulsion is involved in the relation between them – any more than it is in the relation between points and the plane that they belong to.

Determinism in the sense in which it is most often understood in public debate does seem to involve that outside compulsion. It is only made to look plausible by confusing those two senses. The mixed doctrine that results is not workable because it involves fatalism – a belief that all conscious effort is futile – which is not a view that anybody could actually live by, least of all anybody who goes to the trouble of forming arguments. As we shall see, if deterministic and mechanistic sages really believed that conscious effort had no effect, they would not take the trouble to write their books. Fatalistic people do not in fact write such books, because it is hard work to do so.

The concept of responsibility which is built into science, as it is into the rest of scholarship, bears out this unfortunate truth. People are not supposed to get Nobel prizes for work which they did not themselves consciously attend to, and this is no accident. What we are honoured and blamed for is our conscious effort. Honours and criticism alike attend work done on purpose by the whole person, not work ghosted by someone else – including one's neurones – while one is asleep. It is the whole person who is honoured or criticised, because the work is invariably taken to be done by that person. Of course this person needs and uses suitable neurones, but that is another matter.

THE WIDER CONTEXT

The concepts of responsibility and freedom which emerge over free will bring us back in the third part of the book to the social aspects of our identity – to questions about how the kind of self that we are dealing with fits into the wider world and to the inadequacies of social atomism in dealing with this.

Any realistic notion of ourselves rests on the recognition that we ourselves – weak, ignorant and transient though we are – are certainly responsible beings, not bits of helpless dust floating in the wind. Responsibility, however, is the condition of a social creature, not of a stone or a solipsist. It is always responsibility to and for those around us.

This 'whole person' of whom we have been talking is not, then, a solitary, self-sufficient unit. It belongs essentially within a larger whole, indeed within an interlocking pattern formed by a great range of such wholes. These wider systems are not an alien interference with its identity. They are its home, its native climate, the soil from which it grows, the atmosphere which it needs in order to breathe. Their unimaginable richness is what makes up the meaning of our lives. The self's wholeness is not, then, the wholeness of a billiard-ball but that of an organism, a transient, struggling creature which has, of course, its own distinct shape but which still belongs in its own context and background. Much though this being values its own freedom, it unavoidably looks for its fulfilment to horizons far beyond its private destiny. The third part of this book asks how far these horizons extend – what place we take in a wider whole – what range of wider claims it is in our nature to recognise?

During the last four centuries political thinkers in the West have concentrated mainly on limiting those claims. They have put genuinely heroic efforts into cutting bonds. They have managed to free people from endless forms of oppression, both political and domestic, and of course this has been a splendid achievement. The difficulty is just in seeing what it leads to now.

Freedom itself is a negative ideal. Its meaning depends in each case on what particular bonds it frees us from. The reformers who fought each special kind of oppression were always led by a vision of a particular kind of freedom that would replace it, a special way in which society would be changed when they had cut a certain kind of bond. But it has gradually become plain that this bond-cutting sequence is cumulative, which means that it cannot go on for ever. Humans are bond-forming animals. When all the bonds are cut – when the various kinds of freedom are all added together – when a general vision of abstract freedom from every commitment replaces the more limited aims – then, it seems, we might be left with a meaningless life. It begins to seem doubtful whether any kind of human society is then possible at all.

De Tocqueville, who was immensely impressed by the earlier stages of this emancipating process in America and who certainly wished that burgeoning democracy well, was yet alarmed by psychological consequences which he saw following on from this development – so alarmed that he invented a new name for them: Individualism. People (he wrote) were beginning to feel that

> they acquire the habit of always considering themselves as standing alone, and *they are apt to imagine that their whole destiny is in their own hands* . . . Thus, not only does democracy make every man forget his ancestors, but it hides his descendants and separates his contemporaries from him: it throws him

> back for ever upon himself alone, and threatens in the end to *confine him entirely within the solitude of his own heart.*
> (Alexis de Tocqueville, *Democracy In America*, first published 1835, part 2, book 2, chapter 27; emphases mine)

This diagnosis will be discussed further in Chapter 14.

WHAT ABOUT THE EARTH?

I think that today we are becoming increasingly aware of these dangers, increasingly struck by the limitations of individualism. Of course we still acknowledge its ideals. Indeed, we are still inclined at times to favour any change which can be represented as a freedom, especially a commercial freedom, even if it seems otherwise harmful. But on the whole, the strange tide of social atomism that surged in the 1980s is considerably receding today. We see that we need to rebuild a more realistic attitude to our social nature. But we are somewhat puzzled about how to do it, how to conceive the wider world within which we are now inclined to try and take our place.

One movement in this direction which seems to me really important is the notion of *human rights*, which I discuss in Chapters 15 and 16. In dealing with the distresses of people outside our own nation, we are beginning to free ourselves from the narrow contractual thinking which ruled that these outsiders could not concern us at all. A number of practical considerations are in any case making it clear that globalisation cannot be avoided. Disasters do not respect national boundaries. Ships that sink tend to sink at both ends.

Another consideration, no less important, concerns our relation to the non-human world on which we depend. Since the Renaissance, most sages in our humanistic tradition, both on the right and the left, have neglected questions about that relation. Natural resources were assumed to be inexhaustible. Neither the fear of natural disaster nor the cautious reverence that goes with it in most societies has found much place in official Western thinking since the industrial revolution.

This has really been a bizarre tradition. Its effect is that today, when news continually comes in that these resources are actually failing, we find it simply impossible to take that information seriously. Something beyond the usual unwillingness to accept bad news is surely involved here. The way in which we have been accustomed to think of ourselves as isolated, cerebral units standing above the natural world blocks our understanding of how deeply and directly what goes wrong with that world can concern us. Here I think that recent propaganda for individualism – most notably the sociobiological literature of 'selfishness' – is still dangerously distorting people's perceptions. A quite different imagery is needed to make us grasp realistically that we are actually part of the natural world. I shall suggest that an excellent corrective here is the concept of Gaia – of the world as a self-maintaining whole, comparable to a single organism – a whole within which we, like all other creatures, are involved and play our part.

Thirty years ago, when James Lovelock first displayed this idea, the extreme reductivism then prevailing in biology made orthodox scientists reject it outright. Since that time, however, as the details of the idea have been worked out, a good deal of the science involved has been found to be quite plausible and is now being discussed at a normal level. The main difficulty, however, was never about those scientific details. It concerned the imagery, the vision of a wider whole, in some sense a living whole, of which we are a part. It became clear how much this imagery mattered to the scientists when Lovelock introduced a slightly different image, namely the medical model of the earth as a sick planet needing our care and attention – needing, in fact, a science of *geophysiology* to study its health and sickness.

This way of talking greatly reassured the scientific public. It has allowed detailed discussion of the implications to go on without alarm and outrage. This was reasonable because the metaphor of illness is indeed a valuable one, leaving room for the sense of urgency that we so badly need. By showing the earth as a whole rather than as a loose heap of replaceable resources, it makes it possible for us to see it as vulnerable, capable of health or sickness, capable of real injury. And these are indeed the terms in which we need to think about the alarming changes it is undergoing today.

Yet it is interesting to ask just what is the difference between this image and the original one of Gaia the earth-goddess. Why is one of them acceptable while the other still causes alarm? After all, the medical model too accepts that the earth should be treated as if it were a living whole, which was what was originally supposed to be objectionable.

Is the disturbing factor about the image of Gaia, then, its apparent connection with religion? It is interesting to note how readily such connections are now accepted in the prestigious area of theoretical physics, where topics such as 'the mind of God' are constantly discussed and where few people are keen to question Einstein's dictum that the roots of science lie in religion.

Obviously the idea of Gaia is a myth, a symbol. But then so is the sociobiological idea of the Selfish Gene. One of these myths emphasises our separateness from the world around us. The other emphasises our profound dependence on it. Since wholes are quite as real as parts, there is no reason in principle why we should have to prefer the first emphasis over the second. The choice between them depends on their relevance to our situation. And given that current situation, there seems to me to be little doubt about which of them we most need to guide our thinking today.

Part I

Visions of Rationality

1

THE SOURCES OF THOUGHT

IS ART A LUXURY?

Is there any connection between poetry and science? Academic specialisation usually divides these topics today so sharply that it is hard to relate them on a single map. But there is one very simple map which does claim to relate them, a map which is worth looking at because it has quite an influence on our thinking. It is the map which the distinguished chemist Peter Atkins draws in the course of arguing that science is *omnicompetent*, that is, able to supply all our intellectual needs. He notes that some people may think we need other forms of thought such as poetry and philosophy as well as science because science cannot deal with the spirit. They are mistaken, he says. These forms add nothing serious to science:

> Although poets may aspire to understanding, their talents are more akin to entertaining self-deception. They may be able to emphasise delights in the world, but they are deluded if they

and their admirers believe that their identification of the delights and their use of poignant language are enough for comprehension. Philosophers too, I am afraid, have contributed to the understanding of the universe little more than poets . . . They have not contributed much that is novel until after novelty has been discovered by scientists . . . While poetry titillates and theology obfuscates, science liberates.

(From 'The Limitless Power of Science' in *Nature's Imagination*, ed. John Cornwell, Oxford University Press, 1995, p. 123)

Though this view is not usually declared with quite such out-spokenness and tribal belligerence it is actually not a rare one. A lot of people today accept it, or at least can't see good reason why they should not accept it, even if they don't like it. They have a suspicion, welcome or otherwise, that the arts are mere luxuries and science is the only intellectual necessity. It seems to them that science supplies all the facts out of which we build (so to speak) the house of our beliefs. Only after this house is built can we – if we like – sit down inside it, turn on the CD player and listen to some Mozart or read some poetry.

As we shall see, however, this is not how we actually live our lives, still less how we ought to try to live them. Attempts to impose this pattern have distorted the intellectual scene of late from a number of angles. For instance, the idea that science is a separate domain, irrelevant to the arts, has often produced a strange kind of apartheid in the teaching of literature, a convention whereby important and powerful writings get ignored if their subject-matter concerns science, or even the physical world. Thus, criticism of Conrad's sea-stories tends to treat the storms and other natural disasters in them merely as scenery for the human dramas involved, rather than as a central part of their subject-matter. But if Conrad had simply wanted to study human behaviour he could have stayed in Poland. Similarly, H.G. Wells and the whole vigorous science-fiction tradition derived from him were long cold-shouldered out of the literary syllabus and have not yet fully reached it – even though writers like Conrad and Henry James admired Wells deeply and saw the force of his vision. Until quite lately, even *Frankenstein* was ignored. Potent ideas expressed in these writings have thus not been properly faced and criticised in the teaching of literature. These ideas are, of course, often ones about how the science by which we study the physical world relates to the rest of life, which is an extremely important topic. They include a wide range of matters that can help us in trying to understand and face the environmental crisis.

All this means that intending students face a rather bewildering choice. On the one hand they are offered a narrow, somewhat inward-looking approach to literature. On the other, they face a kind of science-teaching which never mentions the social attitudes and background assumptions that influence scientific thought – indeed, one that often views any mention of these topics as vulgar and dangerous. Thus, they may study either the outer or the inner aspect of human life, but must on no account bring the two together.

In fact, despite the efforts of many reformers, Descartes still rules. Mind and body are still held apart. Their division tends to produce a population of one-eyed specialists on both sides, specialists who are mystified by their respective opposite numbers and easily drift into futile warfare. It is surely worth while to take a much harder look at the misleading imaginative picture of the intellectual life which is the source of this habit.

LUCRETIUS AND THE VISION OF ATOMISM

This divisive picture is really very odd, one which does not fit the actual history of thought at all. Rereading Atkins' words lately, I began to think about his remark that poets and philosophers 'have not contributed much that is novel [to the understanding of the universe] until after novelty has been discovered by scientists'. What struck me then was the influence that a single great philosophic poem – Lucretius' *On The Nature Of The Universe* – *De Rerum Natura* – has actually played in the formation of modern Western thought, and especially of Western science.

That poem was the main channel through which the atomic theory of matter reached Renaissance Europe. It was forcibly stated there, all ready to be taken up by the founders of modern physics. Of course it was the Greek atomist philosophers who had invented the theory, and no doubt their work would have reached later thinkers in some form even without Lucretius' poem. But the force and fervour of the poem gave atomism a head start. It rammed the atomists' imaginative vision right home to the hearts of Renaissance readers as well as to their minds. That vision included, not just the atomic theory itself, but also the startling moral conclusions which Epicurus had already drawn from it. In this way it forged a much wider strand in Enlightenment thinking.

For Lucretius did not see atomism primarily as a solution to scientific problems. Following Epicurus, he saw it as something much more central to human life. For him it was a moral crusade – the only way to free mankind from a crushing load of superstition by showing that natural causation was independent of the gods. Human beings, he said, are so ceaselessly tormented by anxiety about natural events that they exhaust themselves in precautions against them that are useless and sometimes horrible, such as human sacrifice:

> They make propitiatory sacrifices, slaughter black cattle and despatch offerings to the Departed Spirits ... As children in blank darkness tremble and start at everything, so we in broad daylight are oppressed at times by fears as baseless as those horrors which children imagine coming upon them in the dark. This dread and darkness of the mind cannot be dispelled by

the sunbeams, the shining shafts of day, but only by an understanding of the outward form and inner workings of nature . . .
How many crimes has religion led people to commit.
(*De Rerum Natura*, trans. R.E. Latham, London, Penguin, 1951,
book 2, lines 50–62, book 1, line 101)

Thus it was Lucretius who launched the notion of science as
primarily a benign kind of weed-killer designed to get rid of
religion, and launched it in great rolling passionate hexameters
which gave it a force it would never have had if it had been
expressed in unemotive prose. His work is visibly the source of
the anti-religious rhetoric that is still used by later imperialistic
champions of science such as Bertrand Russell and Atkins
himself.

WHY VISIONS MATTER

This is not just a debating point for the deplorable war of the two
cultures. The story of the influence that Greek atomistic philosophers had, by way of a Roman poet, on the founding of modern
science is not a meaningless historical accident. It is a prime
example of the way in which our major ideas are generated,
namely, through the imagination. New ideas are new imaginative visions, not just in the sense that they involve particular new
images (such as Kekule's image of the serpent eating its tail)
but in the sense that they involve changes in our larger world-
pictures, in the general way in which we conceive life. These
changes are so general and so vast that they affect the whole
shape of our thinking. That is why something as important as
science could not possibly be an isolated, self-generating
thought-form arising on its own in the way that Atkins suggests.
To picture it as isolated in this way – as a solitary example of
rational thinking, standing out alone against a background of
formless emotion – is to lose sight of its organic connection

with the rest of our life. And that organic connection is just what makes it important.

Changes in world-pictures are not a trivial matter. The mediaeval world-picture was static and God-centred. It called on people to admire the physical cosmos as God's creation, but it viewed that cosmos as something permanently settled on principles that might well not be open to human understanding. By contrast, the atomists showed a physical universe in perpetual flux, a mass of atoms continually whirling around through an infinite space and occasionally combining, by pure chance, to form worlds such as our own. This insistence on the ultimate power of pure chance, which is still such an important principle in today's neo-Darwinist thinking, was thus already central to this early atomist vision.

In principle, this new universe was physically comprehensible because we could learn something about the atomic movements and could thus understand better what was happening to us. But it was not morally comprehensible. It had no meaning. According to Lucretius, the attempt to comprehend the world morally had always been mistaken and was the central source of human misery. In their mistaken belief that they could reach such an understanding, anxious and confused people had taken refuge from their ignorance in superstition:

> in handing over everything to the gods and making everything dependent on their whim ... Poor humanity! to saddle the gods with such responsibilities and throw in a vindictive temper! ... This is not piety, this oft-repeated show of bowing a veiled head before a stone, this bustling to every altar, this deluging of altars with the blood of beasts ... *True piety lies rather in the power to contemplate the universe with a quiet mind.*
> (Book V, lines 1185 and 1194–1203; emphasis mine)

This reference to the possibility of true piety is interesting and we must come back to it. But his main point is a simple one. Instead of this anxious pursuit of bogus social explanations for natural events – instead of these wild speculations about irresponsible gods, people should become calmer and look for physical explanations which, though much slighter, would be reliable so far as they went, and would thus quench their anxiety.

That dynamic and chilling yet ordered world-picture made physical speculation seem possible and indeed necessary. At the Renaissance, moreover, it came together with another picture which had not been available before – namely that of the world as a machine. The invention of real complex machines such as clocks gave the human imagination an immensely powerful piece of new material. Machine-imagery changes the world-view profoundly because machines are by definition under human control. They can in a sense be fully understood because they can be taken to pieces. And if the world is essentially a machine, then it can be taken to pieces too and reassembled more satisfactorily. It was the fusion of these two imaginative visions that made modern science look possible. And it had to look possible before anybody could actually start doing it.

This dependence of detailed thought on entirely non-detailed visions is a central theme of this book. The originating visions are, of course, necessarily vague. When the Greek atomists spoke of the various kinds of atoms as having their own specific movements, they had not the remotest idea of what these movements might be or how anybody could trace them. Though it was central to their position that the movements themselves were fixed, definite and invariable, they could not, in the nature of the case, possibly supply examples. They had to convey their point through the necessarily vague medium of imagery. What they were supplying was much more like a Turner sketch than it was like a photograph, and it was not in the least like an engineer's diagram.

At this imaginative stage, then, they were putting forward a theory *about* exactness – they were envisaging an ideal of exactness comparable with that which we now think of as typical of science. But they had not got anything like an exact theory. At this stage, this kind of vagueness is not a vice, any more than it is a vice in a map of the world that it does not show the details of the small areas within it. It is natural and proper that our detailed thinking arises from imaginative roots. But it is important that we should recognise the nature of these roots – that we should not confuse the ideal of exactness with the actual achievement of it.

Impressive and influential theories like this one do not originally gain their influence by telling us exact facts about the world. It is usually a long time before they can provide any such facts. Actual precision comes much later, if at all. But theories are not half-established facts either. They are *ways of looking at the facts* – pairs of spectacles through which to see the world differently. What makes theories persuasive in the first place is some other quality in their vision, something in them which answers to a wider need. There is always an imaginative appeal involved as well as an intellectual thirst for understanding. Theories always answer a number of different needs, needs which those who are moved by them are not aware of. As they are used and developed, this plurality of power-sources begins to become visible and can result in serious conflicts.

THE MEANING OF DETERMINISM

For example, the determinism which the atomists introduced – the belief in a completely fixed physical order – obviously did not originally owe its appeal to being established as an empirical fact. It is an assumption that goes infinitely beyond any possible evidence, one that is made for the sake of its useful consequences. The way in which it seemed to guarantee the regularity of nature was highly convenient for science. But of course that convenience could not show it to be true. Determinism was not and could not be a conclusion about the world proved by scientific methods. It was an assumption made in order to make the scientific enterprise look, not just plausible so far, but infinitely hopeful. In the modern age, however, that infinite hope became more or less compulsory. When twentieth-century physicists began to question this dogmatic determinism it became obvious that scientists did not view it merely as a dispensable tool but as a matter of faith, a central plank of scientific orthodoxy. Einstein, when he objected to the reasonings of quantum mechanics by insisting that God does not play dice, was talking metaphysics, not physics. Karl Popper, commenting on this, remarks: 'Physical determinism, we might say in retrospect, was *a daydream of omniscience* which seemed to become more real with every advance of physics until it became an apparently inescapable nightmare.' (In 'Of Clouds and Clocks' in his book *Objective Knowledge* (Oxford, Oxford University Press, 1972), p. 222; emphasis mine)

Popper suggests that determinism was really welcomed as much for its flattering view of ourselves as for its soothing account of the world – as much because it declared *us* infinitely capable of knowledge as because it claimed that the world itself was ordered and knowable. While both claims are unprovable, it was (he says) the first claim – our own potential omniscience – rather than the second which was really attracting theorists. It still does.

This bias towards establishing and glorifying our own status is still more obvious over mechanism – the further development of determinism which relies on machine imagery, thus producing the delightful impression that in principle we can copy or rejig all natural objects at will as well as understand them. Thus Julien de la Mettrie, meeting objections to mechanism based on the complexity of natural beings, replied that, with a bit more trouble, machines can imitate every kind of complexity:

If more instruments, more cogwheels, and more springs are

required to register the movements of the planets than to denote the hours: if Vaucanson had to employ more art to produce his flautist than to produce his duck, if he had employed still more energy he might have produced a being with the power of speech ... The human body is a clock, but an immense one and constructed with so much artifice and skill that if the wheel which turns the second-hand should stop, then the minute-hand would still turn and continue on its way.

(*The Man-Machine*, ed. M. Solovine, p. 129)

2

KNOWLEDGE CONSIDERED AS WEED-KILLER

FATALISM AND CONTEMPLATION

At this point, however, the messages from these different visions begin to divide. Lucretius and Epicurus did not promise this kind of complete knowledge at all and took no interest in the technology that might grow from it. Their belief in chance went deeper, making them much more radically sceptical. They had no confidence in any practical attempt to improve human life. Epicurus himself actually despised the pursuit of theoretical knowledge for its own sake as one more distraction from the pursuit of inner peace and warned his followers against it. 'Set your sail, O happy youth,' he cried, 'and flee from every form of education.' Lucretius is indeed more interested in details about atoms, but for him too knowledge itself is not the aim. Knowledge is a means not an end, a means to inward peace, not to improved outward activity. Primarily it is a cure for anxiety, a path to *ataraxia*, peace of mind. It is in these terms that he celebrates Epicurus' achievement:

> When human life lay grovelling in all men's sight, crushed to
> the earth under the dead weight of superstition, ... a man of
> Greece was first to raise mortal eyes in defiance, first to stand
> erect and brave the challenge. Fables of the gods did not crush
> him ... He, first of all men, longed to smash the constraining
> locks of nature's doors ... He ventured far out beyond the
> flaming ramparts of the world and voyaged in mind throughout
> infinity. Returning victorious, he proclaimed to us *what can be
> and what cannot: how the power of each thing is limited* ...
> Therefore superstition in its turn lies crushed beneath his feet,
> and we by his triumph are lifted level with the skies.
>
> (Book I, lines 62–78)

This is splendid stuff. But there is no mention here of any
research programme to follow, no talk of the need to track down
every kind of atom and establish its powers. Epicurus' aim is
simply to show, in principle and *a priori*, 'what can be and what
cannot'. Essentially he merely wants to prove a negative, to show
that we need not fear the gods.

That vision of salvation through science – that hope that scientists can ensure human happiness simply by removing religion –
is still familiar today. It seems worth while to ask how far the
people who now preach it actually share Lucretius' vision, which
their language so often echoes. What the Epicureans were preaching was essentially a fatalistic quietism. They did not think that
human happiness could be increased at all either by political
activity or by the satisfactions of love, or indeed by knowledge
either. Instead, they put their faith in a stern limitation of human
ambition, a concentration on what little is possible to us here
and now. They thought that people who had once fully grasped
that they could not change the world at all, either by sacrificing
to the gods or by any other kind of effort, would cease their
anxious striving, would compose their minds, would be able to
enjoy the satisfactions that life actually gave them in the present,

and would console themselves for their sorrows by admiring the cosmos. Knowing that death would not lead to divine punishment or to a gloomy afterlife, they would no longer fear it. When they died, they would be content simply to dissolve away into their constituent atoms. And, since they knew that natural disasters – lightning, earthquakes, diseases – could not be prevented, they would concentrate on facing these things calmly, instead of dissipating their energy on fruitless efforts to influence non-existent gods.

THE LIMITS OF DETACHMENT

But things have not turned out quite like this. Anxiety is (it seems) a more robust and central part of human life than Lucretius supposed. At least, physical science has not put a stop to it. We modern anxiety-addicts may no longer discharge our tensions by sacrificing to the gods but this does not stop us worrying. We simply exhaust ourselves instead in searching for success, or security, or for better life-styles or for new medicines. No doubt, the general thought that the physical world is orderly and reliable does have some reassuring effect. But it doesn't seem to have given us the humility, the salutary recognition of our own insignificance which Epicurus thought would flow from his vision of the vast, formless impersonal universe. We still live largely in the future rather than in the present. We still plan crowds of incompatible schemes and are continually disappointed. Most of us, whether or not we are scientifically educated, still live on jam tomorrow instead of soberly enjoying what we have while we have it, as Epicurus advised.

In fact, though Epicurean morality has much to recommend it, it does not seem to be compatible, as a whole, with the nature of human striving. Historically, Epicureanism was a faith born of very brutal and unmanageable times – hopeless times such as the last century of the Roman Republic when civil war and corruption made it seem that all effort to improve the world was indeed futile. And there were many such bad times both in the Greco-Roman world and during the Dark Ages that followed it. Epicureanism was therefore valued throughout those eras as a spiritual shelter where people could shut themselves away when necessary and try to forget about outside distresses. When things went a little better and civic activity became possible they tended to prefer Stoicism, which left more scope for social interests and duties.

During the Renaissance, people searching for new ideas found both these pagan philosophies helpful in many ways and used them freely in forming later thought. There is much more of both in our current ideas than we usually notice. But there was then a further special reason for welcoming the Epicurean attack on religion, namely the wars and persecutions that disgraced the name of religion in the sixteenth and seventeenth centuries. Enlightenment philosophers such as Hobbes, Hume and Voltaire who were horrified by those wars and persecutions saw a quite new force in Lucretius' invectives against religion. His exclamation, '*Quantum religio potuit suadere malorum*' ('how many crimes has religion made people commit') expressed their views exactly. I believe that this is why atomism itself was once more seen as having a profound moral significance. That is why its world-picture gained so much force and prevalence among a wide readership – a circle far wider than just the scientists who managed to turn it into a successful physical theory. That is how it became so influential in the modern world.

MAGIC, CONTEMPLATION AND RELIGION

Now, what am I *not* saying? I am not saying that the atomic theory would never have emerged in science if it had not had these particular philosophical and poetic roots in Greece and Rome. I *am* saying that, if the theory had had different roots, it would not have brought with it this particular world-picture, this myth, this drama, this way of accommodating science in the range of human activities, this notion of what it is to have a scientific attitude. (It would no doubt have brought a different one.) In particular, there seems no reason to think that the mere advance of science itself would necessarily have brought with it the Epicureans' undiscriminating, wholesale hostility to everything called religion – their notion that the value of science lay primarily in its power to make people happier by displacing religion from human life.

That idea surely is peculiar because the notion of religion that it involves is such a narrow one. It is absurd to talk as if religion consisted entirely of mindless anxiety, bad cosmology and human sacrifice. Of course the anxious, insatiable business of propitiatory rituals which Lucretius describes does play a large part among the vast range of human proceedings which we call religious. But it plays a large part in the rest of social life too. And in religion it clearly belongs in a special department of that range, namely the department sometimes called magic – but magic in the crude sense of a technology aimed at manipulating the gods. Among the great religions, Buddhism rejects this kind of magic entirely and in Judaism there were strong protests against it as early as the Psalms. It was never incorporated into Christian or Islamic doctrine. The idea of bribing or bullying a deity to change your destiny is foreign to the spirit of these religions. Of course it has often crept into religious practice and often become prominent there – for instance in crude petitionary prayer and the sale of indulgences – because anxiety is such a powerful human motive. But when magic has crept in this way, reforming movements have repeatedly been formed within these religions to get rid of it and to point out that this is not what they are really about.

Pagan religions too of course contained far more admirable things than the kind of low-grade magic that Lucretius deplores.

His own tradition included Aeschylus' tremendous dramas about the nature of justice, Pindar's wonderful hymns to Apollo, Plato's myths, the Eleusinian mysteries and plenty more. And indeed, Lucretius himself furnishes an example of the splendid things that paganism could contain in the great opening passage of his poem *On the Nature of Things*. This is a straightforward hymn to Venus – an invocation of her as the spirit of life, the generous maternal force in nature which fills living things with delight and makes possible the whole admirable world around us. Here is devotion to a force and an ideal which is clearly seen as spiritual as well as physical – devotion of a kind which polytheists often express very nobly towards their deities, because, for them, those deities stand for the central forces at work in their life. When Lucretius mentions the Earth, too, he repeatedly has trouble in restraining himself from openly venerating it as our divine Mother[1] in a way that recalls the current embarrassment of some scientists in handling the concept of Gaia – a point to which we will return in Chapter 17.

But beyond this he shows an intense reverence too for the vast atomic system that he portrays – a system that he sees as essentially ordered and universal although in another way it is chaotic. And it is just this reverence that he wants, above all, to arouse in his readers. He hopes that it will take the place of the useless anxiety that now drives them to make their futile sacrifices, that it will cure them of fearing death and also of entertaining idle ambitions, since it will show that earthly success is hollow, trifling and transient in the perspective of this impersonal but splendid universe.

This is undoubtedly a noble vision. But it does seem to be rather a narrow one. In some ways, the Epicurean ideal is quite close to the Buddhist concept of enlightenment through non-attachment, but it is much more purely negative and asocial. There is no element of Buddhist compassion here, no suggestion of delaying one's own enlightenment to promote the salvation of other sentient beings. Epicureanism offers us a private salvation if we will only respond rationally to the physical universe. But we may want to ask, can the thought of that physical universe alone be expected to produce this degree of philosophic detachment? Can such calm acceptance be expected to follow simply from the findings of physical science?

Our culture now contains many scientists deeply convinced of the atomic structure of matter. They know far more about it than Epicurus ever dreamed of. But it is not clear that this acceptance frees them from the fear of death or even tends to cure them of ambition. It doesn't necessarily make them view Nobel Prizes with lofty contempt as mere earthly trifles. If we want to get a wider perspective within which earthly success really does appear insignificant we will probably need to turn to sages who express a more positive, constructive and generous spiritual vision. Weed-killer alone does not seem to be enough to produce a garden.

THE FLIGHT FROM MEANING

All this raises a most interesting issue about the moral consequences that flow from accepting a physical universe which is vast, impersonal and not shaped to our purposes. The Epicureans, who shared that acceptance with us, tried to draw from it a quite narrow, cautious ethic of unambitious moderation. They could not, however, formulate even that modest ethic without doing what Lucretius forbade and trying to comprehend the world morally – that is, to find a meaning in it. The sense of awe and wonder before the mighty cosmos which they said would make our own concerns seem trivial by comparison emerges as the 'true piety' which Lucretius contrasts with the false piety of superstition. He says it is 'the power to contemplate the universe with a quiet mind'. But anyone who really feels that kind of awe and wonder necessarily sees some meaning in the cosmos, and that recognition naturally expresses itself in religious terms. Lucretius' surge of worship for Venus as the generous, nurturing, maternal principle of life is entirely typical of the thinking that results from it.

Is it possible somehow to avoid this acknowledgement of meaning? Theorists anxious to wipe out all trace of religion have repeatedly tried to reach a pole of total meaninglessness, but their efforts always end by expressing, not a vacuum of meaning but a different meaning, a different drama and not necessarily a better one. Gods are much easier to remove than demons. The picture that most easily results is the exciting, but not very edifying, one of a free fight between *Homo sapiens* and an evil universe. Thus T.H. Huxley shows us a brisk scene which is both Oedipal and military: 'Ethical nature, while born of cosmic nature, is necessarily *at enmity with its parent* . . . The ethical process is in opposition to the principle of the cosmic process . . . Laws and moral precepts are directed to the end of *curbing the cosmic process*'. (T.H. Huxley, *Evolution and Ethics* (Romanes Lecture), Macmillan, London and New York, 1893, pp. viii, 44, 82. Emphasis mine.) How (we might ask) would anyone go about doing that . . .? And how in that case did the ethical process ever get started . . .?

Huxley was trying to protest, quite rightly, against Herbert Spencer's treatment of evolution as a complete guide to ethics. But his own story inflates his protest into a strange kind of animism with the cosmos personified as a bad parent. This kind of projection caught on and in our own age it has – rather surprisingly – been widely treated as scientific, even as a necessary consequence of Darwinism. Thus Steven Weinberg:

> It is almost irresistible for humans to believe that we have some special relation to the universe, that human life is not just a more-or-less *farcical* outcome of a chain of accidents reaching back to the first three minutes, but that we were somehow built in from the beginning ... It is very hard to

realise that this is all just a tiny part of an overwhelmingly *hostile* universe . . . It is even harder to realise that this present universe has evolved from an unspeakably unfamiliar early condition, and faces a future extinction of endless cold or intolerable heat. The more the universe seems comprehensible, the more it also seems *pointless*.

(Steven Weinberg, *The First Three Minutes*, London, Andre Deutsch, 1977, p. 154; emphases mine)

Again, Weinberg's main intention is sound. He wants to attack the idea that the universe exists mainly to produce and cherish *Homo sapiens*. But why should rejecting that foolish view make us rebound to the opposite extreme and personify it as actively attacking and mocking us? This cosmos is, after all, the one that has produced us and has given us everything that we have. In what sense, then, is it *hostile*? Why this drama?

It is worth while to sort out the grievances which make people talk like this. Weinberg accuses the cosmos of being (1) too big, so that it makes us feel small; (2) too temporary, since it is not able to support us and our descendants for ever; (3) too contingent, since it does not show us as necessary; and (4) too pointless, since it does not centrally aim at our interests. This charge-sheet, along with the existentialist atmosphere, comes, of course, from Jacques Monod's attack on 'animism' as the source of all our values:

Science attacks values . . . it subverts every one of the mythical or philosophical ontogenies upon which the animist tradition, from the Australian aborigines to the dialectical materialists, had based morality, values, duties, rights, prohibitions . . . Man must at last wake out of his millennary dream and discover his total solitude, his fundamental isolation . . . He must realise that, like a gypsy, he lives on the boundary of an alien world, a world that is deaf to his music, and as indifferent to his hopes as it is to his sufferings and his crimes. [Values] are his and his alone, but now he is master of them they seem to be dissolving into the uncaring emptiness of space . . .

Life appeared on earth: what, *before the event*, were the chances that this would occur? . . . Its *a priori* probability was virtually zero . . . Immanence is alien to modern science . . . Before [the human species] did appear its chances of doing so were almost non-existent . . . The universe was not pregnant with life nor the biosphere with man. Our number came up in the Monte Carlo game. Is it surprising that, like the person who has just made a million in the casino, we should feel strange and a little unreal?

(Jacques Monod, *Chance and Necessity*, trans. Austryn Wainhouse, London, Fontana, 1974, pp. 160, 165, 137; author's emphasis)

It is very interesting to notice how Monod, after starting to talk about *solitude*, at once drops back, just as Huxley and Weinberg do, into highly emotive social imagery – indeed, into what is surely animism. The cosmos, no longer impersonal, suddenly appears as *indifferent* and *uncaring* in a much stronger sense, like a human who ought to care about us and doesn't. Gypsies, after all, do not live in solitude; they live among people who are more or less hostile to them. Next, this cosmos appears as a croupier, an uncaring functionary from whom we have just happened to win something – namely, all the resources we live on – but with whom we have no kind of personal bond.

All these images are social and the point that they share is a destructive one. They are all aimed at rejecting the most obvious model that strikes people when they think about their relation to the universe around them – the model which has occurred to myth-makers in all cultures – namely, that of parent and child. They are there to counter the image that came so naturally to Lucretius of a maternal, bounteous nature. In order to escape this obvious thought Monod is prepared to go to extraordinary lengths, as is seen in his startling claim that 'the universe was not pregnant with life nor the biosphere with man' – meaning (as he explains) that the prior chances of either of these events occurring were 'virtually zero'.

This claim is flatly contrary to the message of modern biology, which is that the causes of complex things like this must accumulate gradually so that their probability steadily increases. Monod seems to see no middle course between claiming that these events are *necessary* – which would be hard to prove – and treating them as totally contingent in the sense of events in a casino, where each happening is deliberately and artificially disconnected from all its predecessors. But the universe is not a casino, even if the order within it is not a strictly necessary one. There are patterns. Organisms all have their relations and their place in a context. They can exist only within quite special circumstances. The comparison with pregnancy is particularly interesting. After all, a pregnancy too can be interrupted. A woman who is eight months pregnant does not necessarily give birth. She may fall off a cliff next week. But this does not make it sensible to say that her chances of giving birth are virtually zero – that they are the same as those of a sterilised woman, which is what Monod appears to be saying both about the origin of life and the rise of the human species. It seems extraordinary that a distinguished molecular biologist can say something so radically unbiological.

Monod backs this view by a singular doctrine about objectivity:

> The cornerstone of the scientific method is the postulate that nature is objective . . . the systematic denial that 'true' knowledge can be reached by interpreting phenomena in terms of final causes – that is to say, of 'purpose' . . . The ancient covenant is in pieces: man at last knows that he is alone in the

unfeeling immensity of the universe, out of which he emerged only by chance.

(ibid., pp. 30 and 167)

But how could abandoning the idea of cosmic purpose mean that we were unrelated to the other parts of the complex biosphere out of which we have arisen? That we have no kindred with whom we are naturally connected? 'Objective' here seems to mean 'totally disconnected', which is certainly not its normal meaning.

I cannot now go further into the meaning of these singular quasi-religious stories, which I have discussed more fully elsewhere.[2] The point that matters here is that they are modern versions of a powerful but not specially rational vision, derived from the atomists, of the natural world as somehow radically foreign to us and of ourselves as radically foreign to that world – a vision that is still influential in our thinking today. The history of its development, and that of some others closely related to it, will occupy us throughout this book.

THE NEED FOR INTERPRETATION

This emotional estrangement of human life from the cosmos is, of course, only one possible way of reading the Epicurean message. And it is important to concede the part that Lucretius got right. The exaltation of science does indeed have a philosophic point. The modern scientific vision of the vast universe does have enormous grandeur. Contemplation of it certainly can enlarge our mental horizons, distract us from mean preoccupations, raise our aspirations, remind us of wider possibilities. This is a real benefit, for which we should be grateful. The trouble about it is that, once we have this new vision, there are many different interpretations that we can put on it, many different dramas that arise, many directions in which it can lead us. It is quite hard to distinguish among those directions and to map them in a way that lets us navigate reasonably among them.

For an obvious instance, we can respond in many different ways to the miracles of modern physics. Some people see in those miracles only the possibility of more and better weaponry. That is why a high proportion of the world's trained physicists are now engaged on military research. Beyond this, there is a wider circle of people to whom these miracles chiefly mean just an increase in power, without any special idea about how that power had best be used. Out of this fascination with new power there arises our current huge expansion of technology, much of it useful, much not, and the sheer size of it (as we now see) dangerously wasteful of resources. It is hard for us to break out of this circle of increasing needs because our age is remarkably preoccupied with the vision of continually improving means rather than saving ourselves trouble by reflecting on ends. This is the opposite bias from the fatalistic quietism of the Epicureans, who refused to think about means at all. It is not clear that ours is any more sensible than theirs. But, at a casual glance, it is just as natural a response to the grand vision of the physical world presented by modern science.

The difficulty is to make that glance more than casual, to criticise properly the various visions constantly arising in us. We need to compare those visions, to articulate them more clearly, to be aware of changes in them, to think them through so as to see what they commit us to. This is not itself scientific business, though of course scientists need to engage in it. It is necessarily philosophic business (whoever does it) because it involves analysing concepts and attending to the wider structures in which those concepts get their meaning. It starts with the fuller articulation of imaginative visions and moves on later to all kinds of more detailed thought, including scientific thought.

That is why all science grows out of philosophical thinking – out of the criticism of imaginative visions – why it takes that criticism for granted and always continues to need it. It is why the vision of an omnicompetent science – a free-standing, autonomous skill with a monopoly of rationality that does all our thinking for us – is not workable. That idea, which Atkins now somewhat desperately revives, was vigorously promoted by positivistic thinkers from Comte's time on but it does not really make much sense. All science includes philosophic assumptions that can be questioned and those assumptions don't stop being influential just because they have been forgotten. They lie under the floorboards of all intellectual schemes. Like the plumbing, they are really quite complicated, they often conflict, and they can only be ignored so long as we don't happen to notice those conflicts.[3]

When the conflicts get so bad that we do notice them, we need to call in a philosophic plumber – not necessarily a paid philosopher, but someone who knows how the philosophical angle matters. Rationality needs this kind of attention to the conflicts between our various assumptions because rationality itself is something much larger than mere exactness. Rationality is an ideal – one which we perceive somewhat cloudily in a vision, but towards which we can certainly move – an ideal of a just and realistic balance among our various assumptions and ideals.

3

RATIONALITY AND RAINBOWS

THE COGNITIVE ROLE OF POETRY

What, then, about poetry and the arts generally? They too play a central part in our intellectual life because they supply the language in which our imaginative visions are most immediately articulated, the medium through which we usually get our first impression of them. Shelley said that poets are the unacknowledged legislators of the world.[1] This is strong language, but his point may really be needed in an age when literature seems sometimes to have sublimated itself into a haze of texts bombinating above us in a metaphysical stratosphere, and at other times to be viewed simply as a set of rather primitive political documents which must be reproved for failing to reach the most recently invented moral standards. ('Shakespeare was a racist' . . .)

What Shelley meant was, of course, that poets – including, of course, imaginative prose writers – are prophets, not in the sense of foretelling things, but of generating forceful visions. They express, not just feelings, but crucial ideas in a direct, concentrated form that precedes and makes possible their later articulation by the intellect and their influence on our actions. These visions are not something trivial. They are of the first importance in our lives, both when the ideas are good, and, of course, even more so when they are bad. Influential bad ideas need to be understood and resisted so that we can grasp what is wrong with them and replace them by better ones. Thus Primo Levi, describing his early education in fascist Italy, explains how he and his friends took to chemistry largely out of disgust at the ideas presented to them by other studies, simply to get away from

> the stench of Fascist truths which tainted the sky . . . The chem-
> istry and physics on which we fed, besides being nourishments
> vital in themselves, were the antidote to Fascism . . . they were
> clear and distinct and verifiable at every step, and not a tissue of
> lies and emptiness like the radio and the newspapers.
> (*The Periodic Table*, trans. Raymond Rosenthal, London,
> Sphere Books, 1986, p. 42)

This was a reasonable enough way of escape for students when things had gone so far. But earlier, when Fascism had not yet established itself, it needed to be resisted by people who did attend to its ideas. Fascism established itself in Italy largely by means of bad poetry that sold bad visions – by romanticising hatred and violence, by flattering the vanity of voters and distort-ing historical truths. Poets, such as Gabriele d'Annunzio, played a notable part in that campaign. The intellectuals who resisted it never made the mistake of dismissing that contribution as trivial.

Writers – bad and good – who have this kind of effect do not, of course, usually do their legislating by literally spelling out theories, as Lucretius did. They do it by showing forcefully (as novelists and dramatists can) how the new ideas would work in real life. Shakespeare does this all the time, though many of his insights are by now so built in to our thinking that we scarcely notice them. Pope and Blake, Tennyson, Nietzsche and Eliot each powerfully shaped the spirit of their respective ages. And as for the ideal of detachment from worldly affairs which the Epicureans offered, it has taken many forms and suggested many different life styles, which have been displayed imaginatively. Plato presented one of these life styles dramatically in his portrait of Socrates. Nietzsche presented quite another in his *Zarathustra*. Plenty more patterns are available, ranging from the totally ascetic and solitary to the altruistic, patterns which would lead to very different lives. In *War and Peace*, Pierre attempts this kind of detachment and the difficulties he runs into cast a sharp light on its problems. Since we often have to make choices among paths of this kind, literature continually plays a vital part in life by stirring our imaginations and making us more aware of what particular choices can involve. If it is taught in a way which plays down or even suppresses this practical effect, something vital is lost. And we may begin to wonder why, on these conditions, it is important to study literature at all.

It is worth noticing, too, that in this way philosophy itself is a branch of literature. Any major kind of philosophising always presents some distinctive ideal for life as well as for thought because life and thought are not really separate at all. The great philosophers of our tradition have usually displayed their ideals quite explicitly, as have those of other traditions, and if contemporary academic philosophers suppose that they are not doing this they are mistaken. Academic narrowness is a way of life as much as any other. It is quite as easily conveyed by a style of writing, and even more easily by a style of teaching.

THE IMAGINATIVE ROLE OF SCIENCE

All of us who talk and write reveal what kind of things we think important and what we think trivial. We each communicate our particular vision of life, and people who deal in large visions convey them more clearly and more influentially than the rest of us. As I have been pointing out, science too incorporates these influential visions, and this is certainly not something that scientists should avoid or be ashamed of. Scientists have influenced our life profoundly and they do so increasingly today. However much they may try to be objective about particular facts their own personalities inevitably influence the general shape of their message.

This is perfectly legitimate so long as it is done consciously. Scientists influence us by their imagery, by their selection of topics, by the terms in which they explain their theories, by the views that they express about what does and what does not constitute a proper scientific attitude. These things contribute to the constant ferment of dialectic that goes on as we try to compare different attitudes and balance the force of various ideals. What is needed is simply that they know that they are indeed making this kind of contribution – that they listen to the other contributors and grasp the general shape of the debate as Galileo and Darwin and Huxley did. What they need to avoid is fundamentalism – the conviction that the particular imaginative vision espoused by their own party of current scientists is a solitary gospel which must always prevail.

It is, however, very hard for scientists to make this distinction today – hard to detach themselves sufficiently from their favoured vision to see it as one among others in its historical context. Thus Richard Dawkins, who rightly takes poetry a good deal more seriously than Atkins does, has written a whole book (*Unweaving the Rainbow*[2]) to correct the errors of those poets who have attacked Western science with alarm as something dangerous to humane ideals. He notices that all the great romantic poets did at one time or another make powerful statements to this effect. Thus Blake wrote

> May God us keep
> From Single vision and Newton's sleep!
> > (Poem to Thomas Butts, 'With happiness stretch'd
> > across the hills', lines 87–8)

and Keats 'believed that Newton had destroyed all the poetry of the rainbow by reducing it to the prismatic colours' (*Unweaving the Rainbow*, p. x).

I must apologise, incidentally, for repeatedly choosing my quotations from Dawkins. I do not do this in order to persecute him, but for a reason that is greatly to his credit, namely, because he writes so clearly. Clear expressions of important mistakes are very useful things, making it much easier to move on beyond those mistakes than it is when they are wrapped in confusion. In the present case it is also greatly to Dawkins' credit that he does not, like Atkins, simply consign these poets' works to the flames. Instead, he quotes some of them and then remarks kindly what a pity it is that they were so misled. Noting that 'Blake did not love science, even feared and despised it:

> For Bacon and Newton, sheath'd in dismal steel, their terrors hang
> Like iron scourges over Albion: Reasonings like vast serpents
> Infold around my limbs . . .
> > (William Blake, 'Bacon, Newton and Locke' in *Jerusalem*, 1804–20)

Dawkins merely comments: 'What a waste of poetic talent. . . . It is so sad to think what these complainers and nay-sayers are missing . . . It is my thesis that poets could better use the inspiration provided by science . . . we need to reclaim for real science that style of awed wonder that moved mystics like Blake.'[3]

THE BACONIAN ONSLAUGHT ON NATURE

It was not, however, the Romantics who invented this alarming picture of science as a crude and aggressive conqueror. That picture came from the first champions and popularisers of modern science themselves – from the men of the Royal Society and above all from that arch-populariser and cultural hero of the movement, Francis Bacon.

It is notorious that Bacon regularly described scientific activity in oddly savage imagery, incorporating violent conquest as a central part of his original myth of scientific supremacy. This rhetoric has shaped later conceptions of science profoundly. Bacon repeatedly insisted that the aim of the new science must not be just to 'exert a gentle guidance over Nature's course' but 'to conquer and subdue her, to shake her to her foundations'. Scientists must do this in order to be able to turn 'with united forces against the Nature of Things, to storm and occupy her castles and strongholds and extend the bounds of human empire, as far as God Almighty in his goodness shall permit' – that is, as far as was physically possible. That victory would inaugurate something that he strangely called a 'truly *masculine* birth of time', a new epoch which would subdue 'Nature with all her children, *to bind her to your service and make her your slave*'.[4]

Bacon's official point in using this violent imagery was to insist on the importance of experimental methods in science. He described this as 'putting nature to the question' – that is, in the language of the day, putting her to the torture in order to force her to answer. Such forcing obviously seemed to him to involve violent conquest. We may well wonder why he felt that this particular drama was necessary in order to make his point about method.

Knowledge, said Bacon, is power. And of course, so it is. But if power is your main aim in seeking knowledge, then you become liable to all the distortions of thought that notoriously afflict power-seekers. Thus, in Bacon's *New Atlantis*, the Director of the research-institute called Salomon's House naively declares that the 'End of our Foundation is the Knowledge of Causes and secret motions of all things, and the *enlarging of the bounds of human empire, to the effecting of all things possible*'.[5] But what sort of an aim is that? Clearly, most possible things would not be desirable at all. Many of them would be pointless, many would be disastrous, many would contradict one another. What would it mean to aim at effecting all of them? These strange passages express a childish intoxication with the idea of unlimited power itself rather than any sane notion of science. It is the same intoxication which, today, still suggests the idea that everything that becomes possible to technology ought, for some reason, actually to be done.

ATTRACTION, GRAVITATION AND LOVE

We need to understand the sources of this strange language. In calling for a hostile attitude to nature, Bacon was not just urging scientists to drop a casual folk tradition of treating it with respect. He was taking sides in a current dispute within science itself.

During the Renaissance, the idea of explaining natural phenomena in terms of sympathies and attractions between various substances within a wider natural system was prevalent among perfectly serious students as well as among sorcerers. These students used the name 'magic' in an innocuous, neutral sense for those forces, not implying that they were supernatural but merely that they were mysterious, that their working could not be fully understood. Primarily this meant that they could not – as mechanistic thinkers proposed – be explained entirely in terms of impact. Studying these forces involved looking at natural systems as larger wholes rather than just as chance assemblies of parts. It also naturally suggested a reverence, gratitude and respect for the still wider, still more mysterious system of nature that included them all. Equally naturally, it gave rise to the thought that the scientist himself should try to operate as a harmonious element within that enclosing system rather than as something alien attacking it. As Marsilio Ficino put it:

> All the power of magic consists in love. The work of magic is the attraction of one thing by another in virtue of their natural sympathy. The parts of the world, like the members of one animal . . . are united among themselves in the community of a single nature. From their communal relationship a common love is born and from this love a common attraction, and this is the true magic . . . Thus the lodestone attracts iron, amber, straw, brimstone, fire, the Sun draws leaves and flowers towards itself, the Moon the seas . . .
>
> (Marsilio Ficino, *Commentaire sur le Banquet de Platon*, trans. R. Marcel, Paris, 1956. Quoted by Brian Easlea, *Witch-Hunting, Magic and the New Philosophy*, Brighton, Harvester Press, 1980, p. 94)

From this perspective it seemed perfectly appropriate for scientists to love and revere nature and to approach it sympathetically rather than otherwise. It also seemed fitting to continue to use – as Lucretius did – the traditional imagery that spoke of it as female, a bounteous mother. It must be repeated that this way of thinking was still not in itself superstitious (though it could obviously be used superstitiously) since the forces involved were seen as entirely natural, not as supernatural or occult. Johannes Kepler, who took this idea of natural magic very seriously, was following it when he proposed his theory of gravitational attraction and in particular when he used it to explain the tides.

Galileo, however, rejected that explanation and the whole idea of gravitational attraction that went with it because he was committed to the rival mechanistic model which saw the cosmos as a collection of separate particles interacting only by collision. As Descartes put it:

'There exist no occult forces in stones or plants. There are no amazing and marvellous sympathies and antipathies, in fact *there exists nothing in the whole of nature which cannot be explained in terms of purely corporeal causes totally devoid of mind and thought.*'[6]

Since the mechanists held that particles must remain radically separate, the whole possibility of electric connections between them would have been unthinkable for them, though not for Kepler. This is why Galileo could not accept Kepler's two vital discoveries, of the elliptical orbits of the planets and the moon's influence on the tides, viewing them both as superstitious and irrational.

In the end, of course, mechanistic thinking led to dazzling successes in some areas of science. But its glory was marred for a time by embarrassing failures in other cases where mechanistic methods simply failed to work – notably over gravitation and in the electric and magnetic examples cited by Ficino as well as over the development of embryos. Mechanists were extremely uneasy about these awkward examples. Their unease may partly explain the brutal rhetoric with which they rejected any suggestion of a more reverent, holistic view of nature which might claim to explain them better.

Accordingly, when Newton first proposed his doctrine of gravitational attraction, the dominant mechanistic consensus of his day rejected it as incomprehensible. He was only able to get it accepted eventually by suggesting tactfully that what made this mysterious attraction possible was a divine miracle. Absolute space was God's sensorium, and its unifying force sufficed to bring together the essentially separate objects within it. So great a miracle as this could properly be viewed as one more proof of the existence and power of God, so that the doctrine could now become highly acceptable to the orthodox. Thus, in explaining his theory to Richard Bentley, Newton began his letter by declaring, 'When I wrote my treatise about our system [the Principia] I had an eye upon such principles as might work with considering men for the belief of a Deity: and nothing can rejoice me more than to find it useful for that purpose.' The proof, however, would only be valid if it was really impossible for attraction to work by non-miraculous means. Newton therefore went on to make this plain.

> It is inconceivable that inanimate brute matter should, without the mediation of something which is not material, operate upon and affect other matter without mutual contact, as it must do if gravitation, in the sense of Epicurus, be essential and inherent in it. And this is one reason why I desired you would not ascribe innate gravity to me.[7]

Thus, as Brian Easlea puts it, 'not only had Newton produced an excellent account of terrestrial and celestial phenomena, but that account could be used, and most certainly was used by Newtonian theologians, to argue the existence of God and thus to underwrite social stability and thereby the established social stratification'.

THE FEUD CONTINUES

What, however, is the difference between miracle and magic? Is it simply a matter of the particular deity involved? Is it perhaps even a matter of the gender of that deity? An interesting change of myth does seem to have been occurring here, as the Sky God – whom many cultures see as loving and working with the Earth Goddess – began to be seen as finally rejecting her, viewing her as a dangerous rival and insisting that he must reign alone. The imagery of a female Nature, a force working under God, which had been largely accepted or at least tolerated by earlier Christian sages, now became blacklisted as hostile both to faith and science. We will come back to that disturbing issue in Chapter 18.

Meanwhile, in tracing the strange history of these disputes we need to note that, even after Newton's dubious reconciliation, the bitterness persisted, and it still persists today. It has done a great deal of damage to our wider culture, particularly to the status of science itself, because the conflict between opposite emotional attitudes to nature became merged with the wider polarisation that viewed Feeling and Reason as rival principles waging a war within human life – a war which, until the mid-eighteenth century, Reason was confidently expected to win.

In this war, a very interesting asymmetry about rhetoric emerged. Highly emotive Baconian hostile imagery about nature was counted as *scientific* and therefore belonging to the language of reason, while affectionate and respectful imagery such as Ficino used was dismissed as mere sentiment. This ruling still persists and accounts for the fact that no Western scientist after Kepler's time could ever dare to use terms such as *love* for forces of attraction, though today equally anthropomorphic – but hostile – words such as *spite, cheat, selfish* and *grudging* are the accepted coin of sociobiological discourse.

The rise and long survival of this feud between two ways of regarding nature has been a serious and lasting misfortune. Our histories of science mostly suppress it because they ignore the gap between Kepler and Galileo, mentioning only the scientifically usable parts of both their theories which were combined under Newton. But the usable parts of theories are not the only ones that have influence. The question of emotional attitudes to nature is deeply relevant to the way people regard science. Galileo showed how important he thought it by his dismissal of Kepler's work, a dismissal which was certainly not based on scientific grounds. Kepler enthusiastically accepted Galileo's defence of heliocentrism, congratulated him on it and sent his own work on gravity as a helpful supplement to it – which it would certainly have been, notably about the tides. But Galileo simply ignored it, preferring to press on with his own unworkable theory of tides rather than have any dealings with the notion of attraction.

In this way the conception of nature as an enemy to be crushed by an embattled, all-conquering, anti-natural science was built into the idea of scientific 'omnicompetence' from the beginning and it worked there vigorously throughout the Enlightenment. Though Newton himself did not develop its rhetoric, his colleagues in the Royal Society did, and their propaganda identified it closely with the Newtonian system. Thus Robert Boyle complained that

> 'men are taught and wont to attribute stupendous unaccountable effects to sympathy, antipathy, *fuga vacui*, substantial forms and especially to a certain being ... which they call nature, for this is represented as a kind of goddess, whose power may be little less than boundless'
>
> (*The Works of The Honourable Robert Boyle*, ed. T. Birch, London 1722, vol. 5, p. 532)

And again, 'the veneration wherewith men are imbued for what they call nature has been a discouraging impediment to the empire of man over the inferior creatures of God' (ibid., vol. 5, p. 165).

He vigorously promoted this anti-nature campaign by becoming Governor of the New England Company and encouraging the colonists to civilise that territory as quickly as possible. Similarly Henry Power, celebrating the Royal Society in 1664, concentrated chiefly on its destructive achievements and on its future supremacy:

> Methinks I see how all the old rubbish must be thrown away, and the rotten buildings be overthrown, and carried away with so powerful an inundation. These are the days that must lay the foundation of a more magnificent Philosophy, never to be overthrown ... a true and permanent Philosophy.
>
> (Henry Power, *Experimental Philosophy*, 1664, London, Johnson Reprint, 1966. Introduction by M. Boas Hall, p. 192)

4

THE SHAPE OF DISILLUSION

THE POPULACE GROWS UNEASY

We will come back to this question about respectful and hostile attitudes to nature in Chapter 17. Meanwhile, we have reached the point where we can see why this identification of science by its self-proclaimed champions with an imperialistic attack on nature began, after a time, to cause real alarm, not just about new technology, but about science itself. When knowledge and power are so closely identified, the suspicion naturally arises that (as John Ziman puts it) 'the evil factor is knowledge itself: science is characterised as a materialistic, anti-human force, a Frankenstein monster out of control'.[1] Keats and the other romantic poets who protested against the dominant mechanistic ideology were not being irresponsible or insensitive to the beauties of science. They did not need to celebrate those beauties because, by the time they arrived on the scene, that celebration had been going on with all stops pulled out for the best part of a century. Newton had long been adopted as the patron saint of the Age of Reason. His name was used, just as those of Marx and Freud, Adam Smith and Darwin are used in our own day, to stand for a wide-reaching ideology that went far beyond his writings – an ideology that was, for a time, dominant and uncritically celebrated. Taking their cue from Pope's epitaph for Newton himself:

> Nature and Nature's laws lay hid in night,
> God said 'Let Newton be!' and all was light –

whole strings of bards proclaimed the wonders of science, whether in blank verse or, more often, in heroic couplets. Newton's skilful concordat with the Creator (and miraculous preserver) of the cosmos ensured that these celebrations were accepted as entirely orthodox and respectable from a religious standpoint, as well as instructive. Thus the official torch of scientific song was passed down without faltering to Erasmus Darwin (1731–1802) who was still eloquent, for instance on chemical themes:

> Hence orient Nitre owes its sparking birth,
> And with prismatic crystals gems the earth,
> O'er tottering domes the filmy foliage crawls,
> Or frosts with branching plumes the mould'ring walls:
> As woos Azotic Gas the virgin Air,
> And veils in crimson clouds the yielding fair.
>
> (From *The Economy of Vegetation*)

Dawkins, perhaps rather ungratefully, does not much care for Erasmus' verses, saying that they 'do not enhance the science'. Instead, he wishes that the great Romantics had chosen their themes better. Keats and Yeats, he says, would have improved their performance by drawing topics from science rather than 'finding solace in an antiseptic world of classical myth . . . Were these great poets as well served as they could have been by their sources of inspiration? Did prejudice against reason weigh down the wings of poetry?' As for Wordsworth, Dawkins thinks he ought to have said more about how rainbows actually worked: 'If Wordsworth had realised all this, he might have improved upon, "My heart leaps up when I behold / A rainbow in the sky" (although I have to say it would be hard to improve upon the lines that follow)' (*Unweaving the Rainbow*, pp. 27 and 47).

But these are unavailing regrets. At the end of the eighteenth century, this was not an available option. The choice then lay between complacent traditional writers like Erasmus Darwin, writing within Baconian scientism, and the great Romantics doing something totally different – something which was, by then, very badly needed. The scientific celebration had been a long and enjoyable party. But the business of poets and other prophets is not only to celebrate things, and it is certainly not to go on always celebrating the same things. Just as often, they need to denounce things, to shake us from our dogmatic slumbers, to warn us, to point to what is going wrong. Sometimes, that is, they have to act as unacknowledged legislators of the world.

WHICH WAR? WHICH ENEMY?

Scientistic ideology has provoked criticism from various different quarters which are not always sufficiently distinguished today. It is not uncommon still to hear it suggested that the only opposition to 'science' comes either from ignorance or from religion. But in the last three centuries the forces which scientific troops have seen as opposing them have twice been entirely transformed. This change is easy to overlook because the rhetoric of those claiming to defend science has remained surprisingly unchanged.

In the seventeenth century, Bacon and the men of the Royal Society had not the slightest wish to attack orthodox Christianity.

They were mostly devout men who saw their science as closely allied with the Church in resisting unorthodox thinkers such as Kepler who believed in nature and 'natural magic', posited mysterious physical forces at work in the universe, and also often held seditious views on politics. Later, during the nineteenth century, as Enlightenment thought became more secular, the perspective changed. It was then that Christianity itself began to be named as the main enemy of science. And today, though that anti-religious pattern still persists, the 'two cultures' war is conceived in yet another way, as opposing science to the disciplines of the humanities – that is, to studies such as history and philosophy which were formerly seen as its allies, and perhaps also to poetry. Meanwhile parapsychology, which was regarded as perfectly scientific in the nineteenth century, has been conscripted as an extra enemy alongside religion.

Since these different supposed conflicts concern different areas of life and thought, they clearly involve different conceptions of *what science itself is* – different conceptions of why it is excellent and of what it is trying to achieve. The 'science' that excluded Kepler's doctrine of gravitation and enthusiastically accepted theism cannot be the same thing as today's science which reverses those positions. And these various conceptions cannot help having a social meaning. To take sides for or against various elements in society in these various wars is inevitably to choose a particular notion of that society. To oppose some particular current ideology is to espouse and express a different one – an alternative idea of life as a whole, a distinctive view of what is important in it.

Such opposition cannot just be something internal to science. It is not 'value-free'. Asking for more science and less of something else is itself a social and political move. This move can be quite legitimate but it must not be mistaken for a part of a pure, mysteriously objective science which stands outside society. Past changes should surely make us think carefully about why we are now inclined to think of particular attitudes as demands of science.

That is why the history and sociology of science are not luxuries but essential tools for any attempt to grasp its role in our lives.

OBJECTIVITY, FEELING AND GENDER

The fear of science which has been expressed from these various angles – the fear that science may act as 'a materialistic, anti-human force' – is not, then, a gratuitous fantasy. It has been a natural response to certain powerful ideas which have long been associated with Western science because they were genuinely professed and linked with it by its early champions – ideas which are still influential and have not yet been explicitly enough disowned. For instance, the association of the notion of science with crazy and irresponsible power-fantasies is still constantly illustrated by a mass of crude science fiction and also by a good deal of actual technology, notably in weaponry. But at the time of the Romantic Revival what discredited it most directly was its association with an attitude of fear and contempt for the imagination and for ordinary human feeling.

This association is still alive and well in our own time. For instance, the behaviourist psychologist John B. Watson advised parents, in a widely read psychological handbook in the 1930s, not to hug or kiss their children because such conduct was not 'objective', and speculated that it would be 'more scientific' not to allow parents – who were clearly prone to this behaviour – to bring up their own children.[2] In the seventeenth century, in the founding days of modern science, the notion of scientific reason as triumphantly opposed to both feeling and fancy was central and was constantly dramatised in terms of gender, thus making reason the exclusive mark of Man and stigmatising feeling as a female weakness. This gendered opposition between intellect and feeling was the point of Bacon's claim about a 'masculine birth of time' and of Henry Oldenburg's boast that the Royal Society would 'raise a Masculine Philosophy'.[3] His colleague, Joseph Glanvill, spelt the point out still more plainly:

> Where the Will, or Passion, hath the casting voice, the case of Truth is desperate ... And yet this is the miserable disorder into which we are lapsed ... The Woman in us, still prosecutes a Deceit like that begun in the garden, and our understandings are wedded to an Eve, as fatal as the Mother of our miseries. And while things are judged according to their suitableness, or disagreement to the Gusto of the fond Feminine, we shall be as far from the tree of knowledge as from that guarded by the Cherubim.
>
> (*The Vanity of Dogmatising: The Three Versions*, ed. S. Medcalf, 1661 Version, Brighton, Harvester Press, 1970, pp. 118, 135)

The natural consequence was that educators in the Age of Reason not only typically ignored the development of the feelings but often tried, so far as possible, to suppress them entirely. Among the many autobiographers who described the effects of that system on their lives, John Stuart Mill put the point well. His father (he writes) was not unkind, but he

> never varied in rating intellectual enjoyments above all others ... For passionate emotions of all sorts, and for everything which has been said or written in exaltation of them, he professed the greatest contempt. 'The intense' was with him a byword of disapprobation ... The element which was chiefly deficient in his moral relation to his children was that of tenderness.
>
> (John Stuart Mill, *Autobiography*, London, Longmans Green and Co., 1908, chapter 3, p. 28)

As Mill explained, this occasioned a chronic social permafrost of which he himself did not become fully aware until, in early manhood, it struck him down with a depressive breakdown that almost drove him to suicide. He began to grasp its meaning, however, on an earlier visit to France where he was somewhat bewildered to notice a different emotional climate:

> I did not then know the way in which, among the ordinary English, the absence of interest in things of an unselfish kind ... and the habit of not speaking to others, nor even much to themselves, about the things in which they do feel interest, causes both their feelings and their intellectual faculties to remain undeveloped ... reducing them, considered as spiritual beings, to a kind of negative existence ... But I even then felt, though without stating it clearly to myself, the contrast between the frank sociability and amiability of French personal intercourse and the English mode of existence, in which everybody acts as if everybody else (with few or no exceptions) was either an enemy or a bore.
>
> (ibid., p. 34)

This careful, self-protective deadening of the feelings under the banner of rationality and science was Keats' target in the lines from *Lamia* that distress Richard Dawkins so much:

> Do not all charms fly
> At the mere touch of cold philosophy?
> There was an awful rainbow once in Heaven:
> We know her woof, her texture: she is given
> In the dull catalogue of common things.
> Philosophy will clip an angel's wings,
> Conquer all mysteries by rule and line,
> Empty the haunted air, and gnomed mine –
> Unweave a rainbow, as it erewhile made

The tender-personed Lamia melt into a shade . . .

(Keats, *Lamia*, part 2, lines 229–39)

If we want to understand this, we had better notice what is happening at that point in the poem. Keats has just told the Greek tale of the mystery woman who is really a snake and who is unmasked as such at her wedding-feast by a philosopher. He then suddenly steps outside the frame and points out something badly wrong with the story itself. He sees it, surely correctly, as anti-life, a piece of propaganda meant as a warning against love and, more particularly, a warning against women. Within the story, Lamia must of course be exposed. People can't marry snakes. But the question is, ought we to frame our life-plans around such stories? Should we expect every woman to be a snake? Should our only reaction to a diamond be to explain that it is just carbon, and to a rainbow to point out that it is just water, as Mill's father would probably have advised? Keats thinks not. And to make the point clear he gives the story a new ending. In his version the deserted bridegroom does not thank the philosopher and rejoice at his escape, as might have been expected. Instead he is desolated and dies of grief.

A hundred years of deliberate one-sided debunking and disenchantment had been enough. The whole spirit of the age was ready to say, with Wordsworth:

Sweet is the lore which Nature brings:
Our meddling intellect
Misshapes the beauteous forms of things: –
We murder to dissect.

(William Wordsworth, *The Tables Turned*)

and the need to say this has certainly not gone away today. People who now – rightly – want to break the association of inhumanity with science cannot just blame the poets for it. They need to acknowledge the influence that inhumane ideas have actually had on the scientific ethos and to get rid of them.

This will not be easy. When Blake wrote of 'Bacon and Newton, sheath'd in dismal steel', his target was, of course, the ideology of mechanism – the eighteenth century Enlightenment's favourite vision of the world as a vast machine which could be put to human use, a machine in which people (other than the theorists themselves) were fairly simple cogs whose working was fully explainable by physical science. That piece of wish-fulfilment is still with us today and it is still used to justify a great deal of folly, not least about computers.

FALSE ANTITHESES

Obviously, what we are dealing with here is not a simple duel between feeling and reason – nor one between science and the arts – to be resolved by a victory for one side. We need somehow to value and celebrate scientific knowledge without being dragooned into accepting propaganda which suggests it is the only thing that matters. Here the story of what happened about optics and the rainbow is particularly interesting, since that example repeatedly served as a symbol for this whole problem. In a most helpful discussion, Stephen Prickett points out that Newton – and more especially his doctrine of optics – 'came to stand for the eighteenth century as a profoundly ambiguous symbol of the whole scientific revolution of which he was only a part'.[4]

In Newton's own day and for some time after it his optical discoveries, especially about rainbows, were indeed celebrated in just the way that Dawkins calls for, as enriching human experience, not impoverishing it. This celebration was entirely harmonious with the orderly spirit of the age and was also, as Pope and Newton himself had suggested, in accord with orthodox religion. It confirmed that confidence in universal order – both cosmic and social – which sustained the Age of Reason. Thus

Mark Akenside in his *Pleasures of the Imagination* (1744) wrote a long eulogy on Newton's insights, beginning:

> Nor ever yet
> The melting rainbow's vernal-tinctured hues
> To me have shone so pleasing, as when first
> The hand of science pointed out the path
> In which the sun-beams gleaming from the west
> Fall on the watery cloud . . .

and James Thomson gave a very similar one in his *Seasons*.[5]

This, however, was only one side of the story. Joseph Addison, though he too rejoiced at the discovery about the rainbow, interpreted it in a disturbing way which undermines the standing of direct experience and gives a much less reassuring cosmos:

> Things would make but a poor appearance to the eye, if we saw them only in their *proper figures and motions* . . . We are everywhere entertained with pleasing shows and apparitions, we discover *imaginary glories* in the heavens and in the earth . . . Our souls are at present delightfully lost and bewildered in a pleasing *delusion*, and we walk about like the enchanted hero of a romance, who sees beautiful castles, woods and meadows . . . but upon the finishing of some secret spell *the fantastic scene breaks up*, and the disconsolate knight finds himself on a barren heath, or in a solitary desert.
>
> (Joseph Addison, *Spectator* 413, 1712)

In short, our experience can only give us illusions. If we want to know reality – the 'proper figures and motions' of things – we can reach it only through science. And if we do that, the feelings with which we now respond to these illusions will be shown up as false and unfounded.

Addison himself does not seem to have minded this idea, since he thought of the illusions as a kindly device of God's, designed to keep us cheerful. But not everybody is as contented as this with a life of chronic deception. As Stephen Prickett points out, it is no accident that the knight at the end of Addison's passage finds himself in exactly the position of Keats' knight in *La Belle Dame Sans Merci*. And this is essentially the same situation as the bridegroom's at the end of *Lamia* – indeed, the two poems fit quite closely together.

Whether Keats read these words of Addison's or not, the bad philosophical reasoning that they expressed was widespread and enormously influential throughout the Age of Reason. It seemed to leave humanity only two choices. Either we could accept an enormous lie – the excitements of normal experience and the feelings that go with it – or, if we rejected that lie, we could face the truth, which was an impersonal, ghostly world of scientific abstractions. That ghostly world would ideally cure us of all emotion, though in view of our natural weakness it might easily plunge us in incurable dejection instead.

This is the simple philosophy that underlay the ideas of a great host of educators, including both Mill's father and Dickens' Mr Gradgrind in *Hard Times*, who wanted nothing taught but facts. It is also the source of Peter Atkins' dismissive view of poetry from which we started, and of many similar muddles today. Its mistakes are, of course, philosophical rather than scientific, which is why Keats, quite rightly, indicts philosophy in *Lamia* and not (as Dawkins suggests) science itself. These mistakes arose largely from the influence of Locke, whose distinction between primary and secondary qualities – between science and perception – was easily interpreted as a simple division between reality and illusion. Bishop Berkeley rightly protested against this:

How sincere a pleasure is it to behold the natural beauties of the earth! ... What treatment then do those philosophers deserve, who would deprive these noble and delightful scenes of all reality? How should those principles be entertained, that lead us to think all the visible beauty of the creation a false imaginary glare?

(George Berkeley, *Three Dialogues Between Hylas and Philonous*, Dialogue 2, ed. G.J. Warnock, Fontana, 1962, pp. 195–7)

We will come back to these difficulties about reality and appearance in Chapters 5 and 12. At present we are concerned with their effect during the Age of Reason, which was that this muddled antithesis of scientific reality versus everyday illusion lined up with a whole constellation of other crude antitheses – science versus literature, intellect versus imagination, analysis versus synthesis, expert versus amateur, man versus woman, adult versus child – thus distorting the whole picture of life and making many of its practical problems look insoluble.

THE BRIDGE-BUILDERS

A great number of people tried to correct this distortion. Their efforts are usually summed up in our histories of thought as forming the Romantic Revival. Not surprisingly, most of these protesters began by simply trying to redress the balance within the existing antitheses rather than rejecting them outright. Indeed this probably had to be done before wider rethinking could begin. At a first glance, this is what Keats seems to be doing in *Lamia* – straightforwardly taking sides with feeling against reason. But clearly his attitude is ambivalent. He is displaying the problems raised by the Addisonian conflict rather than resolving it. He is pointing out that neither alternative is satisfactory. In his letters, he came back repeatedly to that conflict, not to take sides in it but to try to see what was wrong with the terms in which it was presented and how its two sides could be more intelligibly brought together. He was certainly not hostile to science itself.

('Every department of knowledge we see excellent and calcu-
lated towards a great whole. I am so convinced of this that I
am glad at not having given away my medical books, which
I shall again look over.'[6]) Similarly, what Blake objected to was
single vision – the inability to look at things from any angle other
than the scientific one. It was not Newton's discoveries
themselves.

All the great Romantics made this effort to bring both sides
together, which is just what makes them great. Wordsworth and
Coleridge in particular went to great lengths to stress that the
antithesis between thought and feeling was a false one. They
insisted that both were aspects of a single whole that might
best be understood by attending closely to its middle term,
imagination. Here was the scene of the process of creation, both
in art and science – not a mass of idle and delusive fancy, but a
constructive faculty, building experience into visions which
made both feeling and thought effective. A poet, said Wordsworth, had to be 'a man who, being possessed of more than
usual organic sensibility, had also thought long and deeply . . .
Our thoughts . . . are indeed the representatives of all our past
feelings'.[7]

In short, there is no necessary battle between our different
parts; there is merely a great difficulty in seeing ourselves as a
whole. Wordsworth therefore insisted that poetry needs a
skeleton of serious thought. He firmly rejected the anti-
intellectualism of some earlier Romantics. He saw that this con-
ciliatory approach raises the question whether some serious
thought is too technical to be expressed in poetry, in particular,
whether poetry can be expected to deal with physical science.
Wordsworth – himself a great admirer of Newton – clearly
wanted it to do this, just as Dawkins does. What he says about
this project is very interesting. He is so anxious to make the thing
look possible that his discussion becomes fairly tortuous. It turns
on a series of rather puzzling *ifs*.

> If the labours of Men of science should ever create any material revolution, direct or indirect, in our condition, and in the impressions which we habitually receive, the poet will sleep then no more than at present: he will be ready to follow the steps of the Man of science, not only in those general indirect effects, but . . . carrying sensation into the midst of the objects of the science itself. The remotest discoveries of the Chemist, the Botanist or Mineralogist will be as proper objects of the Poet's art as any on which it can be employed, if the time should ever come when these things shall be familiar to us . . . manifestly and palpably material to us as enjoying and suffering beings.

(ibid., p. 939)

He needs that last stipulation because he has not withdrawn his original insistence on the importance of strong feeling, only balanced it by an equal insistence on the importance of thought. The kind of author who would be needed to write this kind of poetry would, then, have to be someone for whom the details of a science are *as familiar and as palpably material* as the topics with which poetry normally deals – topics which are mainly concerned with central human emotions and with the world as we directly perceive it.

Could anyone who was not himself a scientist possibly stand in this relation to the highly technical 'remotest discoveries' of modern science? It surely is not easy to see how Wordsworth's gifted outsiders, eager to follow in the steps of the man of science, could ever reach this kind of familiarity with the details of the subject. They would surely do no better than Erasmus Darwin did, if as well. The whole language of those 'remotest discoveries' has now become completely separate from that of ordinary life, indeed it is often mathematical. 'Sensation' could scarcely be carried into the midst of entities such as quarks and electrons, or even genes.

About these 'remotest discoveries', then, we should perhaps ask Dawkins' question from the other angle. Instead of complaining that poets are not celebrating this aspect of science should we not perhaps rather ask why scientists – who really are familiar with it – are not celebrating it in poetry, and why they never have done so? Some of them do write impressive prose about it, which is a most valuable habit. But it does begin to look as if they may have fairly good reason for not breaking into verse. Lucretius' achievement concerned a very different kind of science and it may not be repeatable today.

On the other hand, the 'general indirect effects' of the sciences – the interface between them and ordinary life – is a topic available to us all. There Wordsworth's 'material revolution' has indeed occurred and it still hits every one of us forcibly. The celebrators of that revolution – Akenhead, Thomson, Addison and the rest – were writing about just that interface. But then so were those who answered them, writers such as Keats, Blake and Wordsworth himself, who protested against the crudity of Baconian ideology. They were not underestimating science. They were discussing the moral and political ideas of its practitioners and their effect on the thought of their age. And since those ideas are still potent today, their protests have not gone out of date. The interface of science and life is not a simple matter and cannot be properly looked at from one side only. 'Single vision' is never adequate.

CONCLUSION

We have seen that Peter Atkins was not mistaken in claiming that science plays a crucial part in our intellectual life. His mistake lies in the somewhat wild suggestion that it occupies the stage alone, that it is the sole contributor of rationality to an otherwise thought-free world. This illusion of omnipotence is strongly encouraged by today's academic specialisation, and it infects other academic tribes as well as the scientific one. We all tend to think that we know best. Historically, today's over-exaltation of science originated as a response to the equally absurd over-exaltation of classical studies that ruled in the nineteenth century. At present, various kinds of rather unreal anti-science rhetoric tend to encourage this warfare by supporting the idea of a simple conflict between a single personified science and some other champion for the single position of intellectual dominance. This idea of a war between two cultures is a futile one. Instead we all need to sit down together and exchange our visions.

By contrast, Dawkins' book *Unweaving the Rainbow* has undoubtedly helped to make this exchange look possible. It is a much more positive contribution to the scene. Dawkins, who really likes poetry, is genuinely distressed to find that on certain issues the poets are not on his side. Unlike many of his colleagues, he really wants to bring these two departments of life together. He asks: why is there this conflict?

And, as we have seen, that is a serious question. It has occupied us throughout the first part of this book and it will crop up again later. We have seen that this is not just a two-party dispute but is part of a whole family of fissures running through our culture in various directions, fissures that have long divided it over matters of the first importance. The rise of modern science did not just increase our knowledge. It also called for a reshaping of our whole conceptual background, a reshaping so difficult that it naturally generated a whole new set of errors and confusions. Even the question of rainbows is far from simple. It is not surprising that the Age of Reason got it wrong. As Gerard Manley Hopkins put it in an early poem:

> It was a hard thing to undo this knot.
> The rainbow shines, but only in the thought
> Of him who looks. Yet not in that alone,
> For who makes rainbows by invention?

And many standing round a waterfall
See one bow each, yet not the same for all,
But each a handsbreadth further than the next.
The sun on falling waters writes the text
Which yet is in the eye or in the thought.
It was a hard thing to undo this knot.

(No. 91 in *The Poems of Gerard Manley Hopkins*,
fourth edition, ed. W.H. Gardner and N.H. Mackenzie,
London, Oxford University Press, 1970)

5

ATOMISTIC VISIONS

The quest for permanence

NO NEED FOR OMNICOMPETENCE

We have seen that science is not and does not need to be 'omni-competent'. It is not an independent, solitary, intellectual citadel, the only scene of rational thought, nor is it a central government under which both poetry and philosophy are minor agencies. The idea of it as thus mysteriously set apart above the rest of life is, however, an important element in our current beliefs. Modern specialisation tends to cut off the physical sciences from the rest of our thinking. What we 'lay people' (as we are significantly called) mostly notice about the sciences is simply their power. Technology impresses us so deeply that we are not much surprised by the claim that scientific methods ought to be extended to cover the rest of our thought. That positivistic claim – first made by Auguste Comte and repeated by many sages since – underlies many desperate attempts today in other studies, especially in the social sciences, to make themselves, in some sense, ever more 'scientific'.

This mysterious segregation of science can, however, just as easily lead to alienation and fear. Both the dangers of technology and the ideological distortions of scientistic thinking lead people to declare war on science itself. Thus we oscillate between idealising science and dreading it.

Both these attitudes are equally wrong. Both deal in unreal abstractions. The sciences (which are many) are not cut off in this way from the rest of our thought but are continuous with it. They don't compose one solid, distinct, autonomous intellectual citadel. The many scientific ways of thinking all grow out of common thought, draw on its imagery and share its motivations. Scientists do indeed aim at objective truth about the world and, like the rest of us, they sometimes achieve it. Water really is made of hydrogen and oxygen and the liver really does secrete bile. But scientists have to select for their investigation patterns which fit patterns in the world they are going to investigate. And the reasons for that selection are by no means always obvious.

I am not suggesting, as some sociologists of science have done, that scientists just make up their results by framing experiments to prove what they already want to believe. Extreme social constructionism is not at all a convincing story. But it surely is striking how deeply scientific thinking is pervaded by patterns drawn from everyday thought and, in particular, how strong an effect the imagery chosen has on what is conceived at a given time as being scientific. To grasp this, let us look more carefully at the meaning and influence of the atomic model.

THE MEANING OF ATOMISM

As we have seen, the pioneers of modern science drew that model from the Greek atomist philosophers and more directly from the passionate Epicurean version of it given by Lucretius.

They accepted it, not just as a scientific hypothesis but as part of a strong and distinctive ideology. They saw it as a symbolic pattern suggesting meanings affecting much wider areas of life. Morally, for instance, atomism seemed to point the way, not only away from religion but also away from communal thinking and towards social atomism – that is, towards individualism. And for scientific knowledge itself, atomism seemed to promise a most reassuring kind of simplicity and finality – a guarantee that the world would prove intelligible in the end in relatively simple terms, once we had split it up into its ultimate elements. In fact, *understanding the world* seemed to be essentially a matter of simplifying it so as to locate those ultimate units. The word *reductivism* is now used to stand for the belief that this kind of reduction is indeed the only, or at least far the best, way of reaching such an understanding.

Both these promises – the social reliance on individualism and the intellectual confidence in final simplicity – were central elements in Enlightenment thinking. Both have been very useful to us and are still prominent in our thought today. But we are now reaching areas where they can no longer help us. On the physical side, scientists no longer think in terms of hard, separate, unchangeable atoms at all but of particles that are essentially interconnected. And, on the social side, attempts to treat people as disconnected social atoms have repeatedly turned out very badly. Yet we still find it very hard to reshape both these thought-patterns. Like a lot of other ideas which we owe to the Enlightenment, they have come to be accepted as necessary parts of rationality. Efforts to change them tend to look like attacks on reason itself. If we want to rethink them as we now need to, I think it will help to glance back and see what made them so appealing in the first place. Readers allergic to metaphysics, or to the history of thought, can probably skip this chapter.

ORIGINS: THE FEAR OF TRANSIENCE

Atomism arose in Greece out of a determined – but ultimately doomed – attempt to find something in the world which was truly fixed and immutable. Attempting to achieve this, Parmenides proposed that this ultimate unchangeable substance must be something entirely hidden, a mysterious whole lying behind changing appearances. The plurality and change that we see around us cannot (he said) be real because they seem to involve a void or nothingness between the changing items, and nothingness is not real. Change and plurality are therefore only illusions and all the things which we actually experience must be unreal. For, as Heraclitus had pointed out, these everyday things are indeed many and are in constant flux, changing as constantly as if they were made of fire. You can never get into the same river twice . . .

Putting the insights of Heraclitus and Parmenides together, it seemed (then) that the whole world around us was indeed unreal. 'Reality' must be a mysterious eternal something lying behind it, a realm which was altogether hidden from us. But this situation struck many people as implausible. Democritus and Leucippus therefore did everybody a service by introducing the idea of atoms. They suggested that the changing world consists of innumerable tiny units which genuinely are changeless – ultimately real in just the way Parmenides demanded – but which really do cause the changes which we see. They called these units *atoms*, which simply means indivisible objects.

An infinite crowd of these atoms, then, swirls around randomly through infinite space and infinite time, colliding and combining now and then by chance to form temporary universes, of which there are many besides our own. There is no purpose anywhere in this process and the gods, who have been formed by it just as much as humans have, do not try to control it. Nor do they interfere in human life. They simply live serenely on their own in the space between the universes. As for mind, it is real enough but it too consists of atoms – very fine, spherical atoms which can move freely through the coarser particles of matter. At death, this mind-stuff dissolves away into its component atoms just like the rest of the body and is lost in the vast cosmic chaos.

We have already seen some of the moral and social consequences of this impressive vision – its usefulness in resisting superstition and also its thinness as a total philosophy for life. Let us look more closely now at its consequences for science and for our notion of what it is to be scientific.

Obviously some of these are good. The sheer vastness of the perspective – the sense of infinite space and time surrounding us – is most impressive. And the central insight that visible processes can sometimes be explained by finding smaller, invisible processes going on inside them is, of course, immensely fertile for the sciences. Yet there is an unbalance in the scheme which is bound to lead to trouble. It concentrates so strongly on the atoms themselves that it has nothing much to say about how they are related.

We naturally ask what forces are making the atoms move. But on this point the atomists were parsimonious to the point of meanness. They thought that nearly all the movement was caused by collisions between the falling atoms. They did not ask what kind of gravity made them fall in the first place. They did add the idea of a *clinamen* or bend – a kind of native, original tendency in the atoms to move slantwise. But that seems only to have been a defence against the objection that otherwise they might fall in parallel like rain and never meet at all. No reason was given for the slant, and since the atoms have no working parts it is hard to see how there could be any such reason. The atoms collided and sometimes got hooked together, but they never truly interacted. Nor, of course, was the slanting motion of any actual use in explaining in any detail why they behaved as they did, still less how they came to be moving in the first place. But more lethally still, even impact itself had not really been explained. The real reason why things such as billiard balls bounce off each other is that the particles at their surface have electrical charges which repel one another. But these Greek-type atoms were not supposed to exercise any force at all. Repulsion was no more available to them than attraction was. They were supposed to be totally inert.

One way and another, then, change had not really been explained at all.

TROUBLE WITH TIME

This difficulty is due to a fact about the original model which has caused lasting trouble – namely, that it is essentially static rather than dynamic. The Greek atomists' notion was that the mere shape and size of the atoms would explain their workings fully without reference to any forces at work or to the kind of whole within which they were working. Their pattern was still that of Parmenides, a timeless pattern requiring an inert whole incapable either of change or relation. The atomists had not got rid of this pattern, they had merely repeated it indefinitely on a smaller scale. Their atoms are tiny Parmenidean universes.

Later developments in physics have not, of course, borne out this insight at all. Since Faraday's time, particle physics has steadily moved away from this static model. Forces and fields are now the main players in the game and mass is interchangeable with energy. Particles are defined in terms of their capacities for action, which naturally vary with the contexts in which they are placed. There is genuine interaction between them. But this scientific development was delayed for a long time by the imaginative grip of the static model – by the belief that impact was indeed the only possible source of movement, the only force that reason could recognise. In particular, as we have seen, the notion of gravitation was long thought to be irrational because it involved action at a distance, not caused by any collision.

Underlying these difficulties, there was a deep and lasting reluctance to admit that change itself could be real at all. Werner Heisenberg, in his profound little book *Physics and Philosophy*[1] remarks on how far modern physics has moved away from this Parmenidean obsession with the static. As he says:

> Modern physics is in some ways extremely near to the doctrines of Heraclitus. If we replace the word 'fire' by the word 'energy' we can almost repeat his statements word for word from our modern point of view. Energy is in fact the substance from which all elementary particles, all atoms and therefore all things are made, and energy is that which moves. Energy is a substance, since its total amount does not change . . . Energy may be called the fundamental cause for all change in the world . . .

> In the philosophy of Democritus the atoms are eternal and indestructible units of matter, they can never be transformed into each other. *With regard to this question modern physics takes a definite stand against the materialism of Democritus and for Plato and the Pythagoreans.* The elementary particles are certainly not eternal: they can actually be transformed into each other.

> (pp. 51 and 59)

As Heisenberg explains, after these collisions the resulting fragments again become elementary particles on their own – protons, neutrons, electrons, mesons – making up their lost mass from their kinetic energy. So if anything can be defined as 'the primary substance of the world' it has to be energy itself. As he puts it: 'The modern interpretation of events has very little resemblance to genuine materialistic philosophy: in fact, one may say that atomic physics has turned science away from the materialistic trend it had during the nineteenth century' (p. 47). Since Heisenberg's time, physicists have begun to say that this primary substance may turn out to be, not exactly energy but some form of space itself. This, however, is not going to be much comfort for materialism.

REALITY AND INTELLIGIBILITY

What *materialism* means here we will consider in a moment. The first thing to notice is that this shift calls for a deep change in our traditional notion of reality, a change which we have certainly not fully made yet. We are free now from the metaphysics which seemed to go with the old physics, from the notion that only the unchanging is real. We don't any longer need to posit static units as the terminus to explanation, treating all explanations as provisional until they reach it.

Change, in fact, is *not* unreal, it is a fundamental aspect of reality. Parmenides thought that changeable, interacting things were unreal because he thought change could not be understood. But this idea flows from a special notion of what it means to *understand* something. Certainly there are some forms of understanding which abstract from time and change, notably in mathematics. And this timelessness does give these explanations a specially satisfying kind of completeness. But for other problems, such as when we want to understand fire or explosions, time and change are part of the subject-matter. And there are other situations again, notably ones involving living organisms, where whole sets of interconnected changes are going on at the same time.

Yet we do gain some understanding of these matters. Thermodynamics and climatology and biology are not just a string of lies and delusions. Their explanations are in a way less complete, less final than those of mathematics, but this is because, being less abstract, they do so much more work. Their greater concreteness allows them to apply more directly to the actual world around us and that world is what we need to explain. That world is (we must insist) not an illusion. It is not a flimsy shell covering a true reality. It is the explanandum. It is the standard from which our notions of reality are drawn. No less a physicist than Richard Feynman celebrates this fact:

> A poet once said, 'The whole universe is in a glass of wine' ... There are the things of physics, the twisting liquid which evaporates according to the wind and weather, the reflections in the glass, and our imagination adds the atoms. The glass is a distillation of the earth's rocks, and in its composition we see the secrets of the universe's age, and the evolution of stars. What strange array of chemicals are in the wine? How did they come to be? ... There in the wine is found the great generalisation: all life is fermentation ... How vivid is the claret, pressing its existence into the consciousness that watches it! If our small minds, for some convenience, divide this glass of wine, this universe, into parts – physics, biology, geology, astronomy, psychology and so forth – remember that nature does not know it! So let us put it all back together, remembering ultimately what it is for. Let it give us one more final pleasure: drink it, and forget it all!
>
> (Richard Feynman, *The Feynman Lectures on Physics*, Reading, Mass., Addison-Wesley Publishing Co., 1963, chapter 3, para. 7)

We need to disentangle the physics here from the metaphysics. The physical question *what stuff (if any) things are made of* is quite distinct from the much more mysterious question of *what reality (if any) lies behind the whole world of experience* – a world of which physics itself is only a part. The ontological question about a presumed reality behind appearance implies a sweeping distrust of *all* experience – including the observations made by scientists. And that distrust needs some special kind of justification.

In the passages just cited, Heisenberg is not just doing physics. He is not just telling us that modern science finds Heraclitus' conceptual scheme more convenient than that of Democritus. He is also pointing out how biased and misleading Democritus' scheme is metaphysically, how it can distort our notion of reality, leading to the notion that mind or consciousness itself is in some sense *not* real. The trouble is not just that Democritus' proposal of fitting mind into the atomic scheme by supplying it with smooth round atoms turned out not to work because there were no such atoms. Even if there had been those atoms, they still would not have furnished a usable way of thinking about mind or consciousness. To do that we have to have a language for the subjective. We have to take seriously what happens at the first-person point of view. And there is no way of doing this inside the atomic scheme, which is irredeemably an external, third-person one.

This is why atomistic thinking led people to metaphysical materialism, to the rather mysterious idea that *only matter is real*. The Greek atomists were the first people who seriously made this striking claim, the first real materialists. Their Ionian predecessors such as Thales had taken it for granted that life and spirit were included as properties of their primal substance – water, air or fire. Instead, the atomists seriously tried to show how life and consciousness could emerge from a world consisting only of static, inert atoms and the void.

This attempt failed resoundingly, and its failure is enormously instructive once we understand it. We are, I think, only now beginning to get it in focus. As Heisenberg says, during the nineteenth century materialism became hugely popular and, in spite of the efforts of modern physicists, on the whole it remains so today. It became an ideology, a creed expressed in a whole stream of devout pronouncements such as that of Karl Vogt that 'The brain secretes thought as the liver secretes bile.'[2] But what do we mean by *reality* if we deny that our own experiences are a part of it? The point can't just be that experience is misleading or unreliable; it has to be that it doesn't happen at all. A brain that has been thinking doesn't deposit any tangible, measurable residue of thought in the experimenter's petri dish. Perhaps, then, conscious thought is just an illusion which should vanish from the equation entirely?

Metaphysical behaviourists such as Watson did sometimes try to take up this startling position but they never managed to make much sense of it. Modern exponents of materialism usually take the more modest line that experience does happen but that it doesn't matter much, that it is somehow *less real*, more superficial than physical processes.

Accounts of events involving consciousness are, then, legitimate at their own level but they are not complete or fundamental. They are only provisional. In order to be made fully intelligible they must be reduced, by way of the biological and chemical accounts, down to the ground floor of physics which is the only fundamental level, the terminus that alone provides true understanding. (This gravitational metaphor is itself extremely powerful and will need our attention.)

EXPLANATIONS

In this form, the materialistic creed doesn't necessarily mention the notion of *reality*, so it is less obviously metaphysical. It appears in more modern guise as a view about explanation and what can make it complete. But of course no explanation ever *is* complete, and, in so far as we do demand completeness, contributions from physics don't necessarily help it. When we ask someone to complete an explanation, what we normally want is something visibly relevant. For instance, if an explanation of a historical phenomenon such as anti-Semitism seems to have gaps in it we ask for material that will fill them. But that material will primarily be historical or psychological because that is what our questions are about. There is no obvious reason why physical details about neurones in the brains of anti-Semites could ever be relevant to the problem. As Richard Feynman explains:

> In order for physics to be useful to other sciences in a theoretical way, other than in the invention of instruments, the science in question must supply to the physicist a description of the object in the physicist's language. They can say 'why does a frog jump?' and the physicist cannot answer. If they tell him what a frog is, that there are so many molecules, there is a nerve here, etc., that is different. If they will tell us, more or less, what the earth and the stars are like, then we can figure it out. In order for a physical theory to be any use, we must know where the atoms are located. In order to understand the chemistry, we must know exactly what atoms are present, for otherwise we cannot analyse it.
>
> (*The Feynman Lectures on Physics*, Reading, Mass., Addison-Wesley Publishing Co., chapter 3, para. 7)

This gap makes a great difference to what we mean by calling an explanation *fundamental*. If we say that a certain explanation has indeed managed to be a fundamental one it won't be because it involves physics. It will be because it answers the central historical and psychological questions that it set out to answer. Some physicalist philosophers believe that in the future, when we know enough about brains, we shall discover quite new kinds of physical explanation which will displace all these existing forms of thought on the matter and will show that they were just superficial 'folk-psychology'. But this is simply a confusion about the kinds of work that different kinds of explanation do. Examining the physical and neurological causes at work in anti-Semitic brains would do nothing at all to explain the ideas involved, any more than examining the brains of mathematicians can explain the mathematics that they are working on. Nobody has yet suggested studying mathematics in this way and it is no more plausible to propose relying on it for explaining the rest of conduct.

POLITICAL USES OF THE MATERIALISTIC VISION

Metaphysical materialism got into European thought in the first place as a weapon used, first by the early atomists and then by political campaigners such as Hobbes, against the dominance of religion. In modern times the prime motivation behind it was horror and indignation at the religious wars and persecutions of the sixteenth and seventeenth centuries and its main target was the notion of the soul as a distinct entity capable of surviving death. As we have seen, this social and political motivation was quite close to that of the ancient atomists, who were also moved by outrage at disastrous religious practices.

This motivation was a suitable one for forging a weapon in campaigns against the churches. But it was much less able to provide a balanced foundation for the whole of science, let alone for a general understanding of life. For that wider understanding, change and interaction needed to be seen as intelligible in their own terms and the first-person aspect of life had to be taken seriously as well as the objective one.

Descartes notoriously saw this last problem and made a magnificent attempt to deal with it by making mind or consciousness the starting-point for his systematic doubt. He did succeed in getting subjectivity finally onto the philosophers' agenda, but for a long time they were puzzled about what to do with it. Descartes still described mind ontologically, not as a first-person aspect or point of view but as a substance, something parallel to physical matter but separate from it and not intelligibly connected with it.

This kind of dualism had the fatal effect of making mind look to many scientists like an extra kind of stuff, not like one aspect (among many) of the real world but like a rival substance competing with matter for the narrow throne of reality. This vision inclined scientifically-minded people to sign up for an ideology called materialism, meaning by that not just allegiance to matter but in some sense disbelief in mind. The idea of the two as rivals for the status of reality persisted. Mind was seen as an awkward non-material entity which ought perhaps to be removed with Occam's Razor, one which was certainly too exotic meanwhile to deserve serious scientific attention. And alarm about it went particularly deep in the social sciences, which were becoming increasingly sensitive about their scientific status.

This is why, through much of the twentieth century, scientists, both social and physical, in English-speaking countries were extraordinarily careful to avoid any mention of subjectivity and particularly of consciousness. In psychology, where this avoidance was fiercely enforced, it was, as we have seen, usually not treated as a metaphysic but as a matter of methodological convenience, since outside behaviour could be observed while inner states could not. But such a choice of method is never likely to be separate from metaphysics. Selection of subject-matter depends on what one thinks important, and judgements about importance are part of one's general vision of the world.

As we have seen, both the method and the metaphysics flow from background presuppositions of which we are often unconscious, presuppositions that are part of our picture of life as a whole. This element in behaviourism became obvious because, before long, strict behaviourist methods were found not to be at all convenient for psychology and had to be abandoned. The attempt to study behaviour without considering the motives behind it could not work because it is not really possible to observe and describe behaviour at all (apart from the very simplest actions) without grasping the motives that it expresses. And since we are social animals, we actually know a great deal about those motives.

It was not convenience, then, that had recommended this method in the first place. The attraction was quite a different one – namely, that it *looked scientific* if one defined scientific method in devoutly materialist terms, in a manner derived from the old atomistic vision, as a method that dispensed with the concept of mind. During the last thirty years, however, notions of mind and consciousness have rather suddenly escaped from the taboo that so long suppressed them. It has been really interesting to watch how they have now become matters of lively debate among a wide variety of academics. Much of what goes on in the now vigorous *Journal of Consciousness Studies* is metaphysics though it is often supposed only to be science. The main point of the enterprise must surely be to forge a new vision that can heal the Cartesian rift between mind and body, showing them, not as warring rivals but as complementary aspects of a larger whole. Physicists like Heisenberg saw the need for that long ago, and we need now to get on with this difficult business.

6

MEMES AND OTHER UNUSUAL LIFE-FORMS

THE SEARCH FOR SOCIAL UNITS

So far, I have been outlining certain rather general ways in which primitive forms of atomism survive and do unrecognised harm to present-day habits of thought. The only case of this that I have yet mentioned is the case of behaviourism in psychology. To show the force of the whole phenomenon we need to notice some others.

Behaviourism itself is quite a striking example because it requires such an unnatural suppression of normal ways of thinking and because, despite this, it had such a powerful influence throughout the social sciences. Indeed, in spite of its official rejection by psychologists, it still remains influential today. Another striking example of quite a different kind does not concern materialism but simply the atomistic habit of breaking up wholes into ultimate units. It is *the social atomism that lies at the heart of individualism* – the idea that human beings are essentially separate items who only come together in groups for contingent reasons of convenience. This is the idea expressed by saying that the state is a logical construction out of its members, or that really there is no such thing as society. A social contract based on calculations of self-interest is then supposed to account for the strange fact that such things as human societies do actually exist.

As I say, this social atomism does not necessarily involve materialism. But the curious pattern which it traces of unrelated human units milling around in a social vacuum echoes very closely the equally strange pattern of unrelated Democritean atoms spinning in the void. The two patterns developed in European thinking at the same time during the seventeenth century and they surely reinforced each other. The physical theory conferred a reassuring scientific flavour on the social one, and the social one, when it began to operate in everyday life, made the physical theory look reassuringly familiar. The language of 'one man one vote' and 'each to count for one and nobody for more than one' sounded not only rational but scientific. At the present day the idea of the Selfish Gene owes much of its appeal to recent revivals of this individualistic pattern. Actual genes are not really individualistic in this way at all: they are elements in a whole within which they need to co-operate quite closely.

The weaknesses of social-contract thinking, and of excessive individualism generally, are now widely acknowledged and perhaps we need not discuss them further here. Michael Frayn sums up its appeal and its weakness admirably in his little book *Constructions*:[1]

> In some moods, at any rate, it seems to us that Robinson Crusoe is the human archetype. Just as philosophers thought that the thick stew of human discourse, with all its lumpy inaccuracies and indigestible assumptions, could in theory be refined down to pure white crystals – atomic propositions embodying atomic fragments of experience – so we feel that

human society, with all its compromises and relativities, is a construction from the series of atomic individuals, each of them sovereign and entire unto himself. We feel that we are Crusoes who have been set down in sight of one another, so that the difficulties of communication and co-operation have been *added* to those of our isolation. As if we are what we are and *then* we enter into relations with the people around us.

But man is the child of man. He comes from the belly of another human creature, seeded there by a third. He can become conscious of his thoughts and feelings only by articulating them in a language developed by communication with his fellows. Even in his inmost nature he is defined by interaction with other beings around him.

(Michael Frayn, *Constructions*, London, Wildwood House, 1964, p. 108)

MEMES

Well, that is one kind of social atomism. Do we want more kinds of it? If we do, we can now have cultural atomism as well, the theory that culture too has an atomic structure, being composed of ultimate units known as memes.

This idea was originally proposed by Richard Dawkins in the last chapter of his book *The Selfish Gene*[2] and has since been taken up by a number of other sages, most recently by the sociobiologist Edward O. Wilson in his book *Consilience*.[3] By positing these units, Wilson hopes to provide a means of reconciling the humanities and social sciences with physical science by bringing them finally within its province. Memes (he says) will form 'the conceptual keystone of the bridge between science and the humanities'.[4]

But is culture the sort of thing that can be understood by dividing it up into ultimate units? It must be, says Wilson, because atomising is the way in which we naturally think. 'The descent to minutissima, the search for ultimate smallness in entities such as electrons, is a driving impulse of Western natural science. It is a kind of instinct.'[5] We need, he says,

> to search for the basic unit of culture ... Such a focus may seem at first contrived and artificial, but it has many worthy precedents. The great success of the natural sciences has been achieved substantially by the reduction of each physical phenomenon to its constituent elements followed by the use of the elements to reconstitute the holistic properties of the phenomenon.

(p. 134)

In fact (says Wilson) it has worked in science so it is surely bound to work in the humanities.

How, then, should it be used? Memeticists differ about this. Wilson himself at first keeps quite close to the pattern set by the discovery of ultimate physical particles. He wants mental *minutissima* – ultimate units of thought comparable to fundamental particles in physics, and he thinks these units can eventually be linked to particular brain-states so as to provide a kind of alphabet for a universal brain-language underlying all thought. At other times, however, he forgets this incredibly ambitious project and describes his particles just as readily as units of culture – obviously a quite different concept.

Other practitioners mostly stick closer to this cultural pattern. Dawkins himself calls his memes 'units of cultural transmission' giving as examples 'tunes, ideas, catch-phrases, clothes-fashions, ways of making pots or building arches' to which he later adds popular songs, stiletto heels, the idea of God and Darwinism[6] – remarkably miscellaneous items which are plainly not the kind of things that could possibly figure as Wilsonian ultimate atoms of thought. Dawkins insists, however, that they are not just convenient, arbitrary divisions either. Indeed, as we saw in the

Introduction, he now considers them not only to be fixed, last-ing, natural units but to be the ultimate units of psychology, superseding any notion of a unitary human individual. Daniel Dennett is equally emphatic about their separate reality. He gives a long list, even more mixed than Dawkins', a list which includes *deconstructionism, evolution by natural selection, impressionism, chess, the Odyssey* and *wearing clothes*, and he insists that these are not conventional divisions but fixed, ultimate units.

> Intuitively we see these as more or less identifiable cultural units, but we can say something more precise about how we draw the boundaries . . . the units are *the smallest elements that replicate themselves with reliability and fecundity.* We can compare them, in this regard, to genes and their components.
>
> (Daniel Dennett, *Darwin's Dangerous Idea*,
> London, Penguin, 1996, p. 344)

ATOMISM MISPLACED

Why then does the *Odyssey* (for instance) contain within it several stories which are well-known in their own right, such as the story of Polyphemus and that of Scylla and Charybdis? *Wearing clothes* is a wide general term covering a vast range of customs. *Deconstructionism* is a loose name used to describe an indefinite jumble of minor theories. And the *idea of God* is also a very wide and ambiguous one. Names of doctrines like these shift their meaning constantly, as indeed does the word *culture* itself, which has undergone many adventures during the past century.

In fact, as Dennett himself points out, most of these items are in a state of constant change, which genes (of course) are not supposed to be. Customs and traditional items of the kind he lists develop constantly, unless we deliberately fix them (as we do the *Odyssey* or *Greensleeves* or chess) by devices such as printing them and publishing rules. This incessant development means that cultural items behave in a way quite like that of whole organisms such as plants or animals but quite unlike the behaviour of genes.

Sometimes, indeed, memeticists do seem to be comparing these cultural items to whole organisms – to phenotypes – and the memes to hidden entities, unseen 'replicators' which are the occult means by which those phenotypes leap from mind to mind. This is the model on which they offer to build a science of memetics, parallel to genetics, as the right means for understanding culture. Memetics is then supposed to instruct us about the devious strategies by which these imagined items go about their business of reproduction, the paths by which they manage to infest us. But it can't do so because there is simply no room for such items. We already know how human beings communicate and influence one another. Occult causes are not needed to explain this process. The pseudo-genes serve no function and need to be cut off with Occam's Razor. From a philosophical angle they are sky-hooks, which is the name that Dennett himself gives to idle metaphysical entities that serve no explanatory function.

Cultural items like these are not, then, ultimate, immutable fundamental particles, atoms of a stuff called culture, nor are they quasi-genes which might transmit it. They are shifting patterns within human life. Understanding them is not a matter of splitting culture into its ultimate particles because culture is not a substance, a solid stuff of the kind which might be expected to consist of particles. Instead, it is a complex of constantly developing patterns, a set of ways in which people behave. And ways of behaving are not the kind of thing which breaks down into ultimate units at all.

Of course it is true that, when we want to understand some aspect of culture such as deconstructionism or the idea of God, we do often analyse it into distinct elements, sub-patterns that compose it. But just as certainly we also need to look for the wider context of ideas out of which these patterns arise, the background which is needed to make sense of them. *Explaining* such things is primarily placing them on a wider map of other ideas and habits, relating the patterns within them to the larger patterns outside. In ordinary life, that is what we do when we want to understand these patterns. And people who want to understand them better – people such as historians, anthropologists, philosophers, social psychologists, novelists, poets and literary critics – have developed, over time, many subtle and skilful ways of carrying this mapping process further.

That kind of cultural mapping is, in fact, the main business of the humanities. The proposal to use the meme pattern as a means of understanding culture cannot therefore serve as a way by which the sciences can liaise with the enormous range of humanistic methods, as Wilson wants. Instead, it is simply a way of ignoring all those methods and offering a meaningless pattern of atomic entities as a substitute. From a scientific point of view, memetics is phlogiston and, what's more, it's unnecessary phlogiston. The idea of phlogiston did at least mark a blank on the map, an empty place which needed to be filled by a proper theory of combustion. But there is no such blank on the humanistic map waiting to be filled by a new, quite general, proposal of this kind for how to start understanding culture. Of course the methods that we now use are imperfect and faulty. But there is no way in which this quite irrelevant suggestion could correct them.

These, of course, are alarming words. Can it (you ask) really be true that these very intelligent, high-minded and highly-qualified people are trying to sell us phlogiston? Can it be true that they themselves have bought it? I fear that it is, and of course I have to explain why I think that such a surprising thing is possible.

The explanation lies in the point that I am trying to make throughout this discussion, namely *the tremendous influence of imaginative visions in general, and of the atomistic vision in particular, in forming our thought-patterns.* This atomistic picture has always had enormous appeal because of its seductive finality. It seems to provide ultimate simplicity and completeness and even a kind of stability, since the atoms at least last for ever even if we don't. When we are trying to understand the shifting chaos of human affairs, the idea of simplifying them in this way is hugely attractive, especially to people who have grown up thinking of the atomic pattern as the archetype of all scientific method, which is the way that all scientists thought in the nineteenth century. Since then the physicists – the original owners of the pattern – have seen reason to drop this seductive vision and to recognise that the world is actually more complex. Many biologists, however, still cling to the atomic model and hope to extend its empire so as to bring order into the muddled rain-forest of human society. But that hope really is mistaken. We need to recognise the atomic vision for what it is – just one possible interpretative pattern among many – and to look for our understandings of human culture elsewhere.

PUTTING MEMES TO WORK: THE WITCH-CRAZE

To show that this dismissal is not arbitrary, it will be best to end this chapter by looking briefly at how the memetic model would work if we did try to use it on a real problem. Dennett, who considers its function more fully than Dawkins, firmly enforces the parallel with 'selfish' genes. It is wrong (he says) to try to explain cultural traits by asking what they do for the people who adopt them. Instead we should recognise that 'a cultural trait may have evolved in the way it has simply because it is advantageous to itself'. A human mind is then

> an artefact created when memes restructure a human brain so as to make it a better habitat for memes . . . Like a mindless

virus, a meme's prospects depend on its design – not its internal design, whatever that may be, but the design it shows the world, its phenotype, the way in which it affects things in its environment [namely] minds and other memes.

(p. 349)

We therefore need memetics to help us grasp the strategies by which memes contrive to infest us even when they are not useful to us; for example: 'the meme for faith, which discourages the exercise of the sort of critical judgement which might decide that the idea of faith was, all things considered, a dangerous idea'. If (then) we want to understand why certain people have faith in somebody or something – for example, if we ask why Western people today tend to believe the declarations of scientists – we should not waste our time asking what reasons or motives they might have for this particular piece of trust. Instead, we should simply note that an entity called *faith* tends to be successful at parasitising human brains.

The next question surely is: how would this approach give us anything that could be called an explanation of human actions? No doubt it is true that it is sometimes hard to find the kind of explanation that we normally look for, namely one in terms of reasons and motives. But we do usually have some idea of those reasons and motives. And if we cannot find them, there is no way in which we can make sense of people's actions. The fact that we sometimes are puzzled in this way is probably what gives the meme idea its only faint plausibility. It is therefore worth while to see how that idea would work if it were used in cases where human motivation really is puzzling.

Consider, for example, the witch-craze which prevailed in Europe from the fifteenth to the seventeenth centuries. That craze was not, as is often supposed, simply a survival of ancient superstition caused by ignorance and eventually cured by science. On the contrary, in the Middle Ages there were few prosecutions for witchcraft. The church authorities did not think it was common and they discouraged witch-hunting because they saw the danger of false accusation. A church canon set strict limits to it. It was in the Renaissance that things changed. At that time, as two recent historians put it:

> The Europeans did three things which set them apart from most other peoples at most times and places. Between 1500 and 1700 they set sail in tall ships and colonised most quarters of the globe. They made stunning strides forward in the sciences. And they executed tens of thousands of people, mostly women, as witches.
>
> (From 'Does Science Persecute Women? The case of the 16th–17th century witch-hunts' by Karen Green and John Bigelow, *Philosophy*, vol. 73, no. 284, April 1998)

This frenzy coincided, then, with the increase of knowledge rather than being cured by it. And, as these authors show, when it finally subsided it did not do so because scientists had shown that witchcraft was impossible but because people gradually began to find it psychologically incredible that there was such an organised host of demon-worshippers.

Here, surely, is something that needs explaining. And I cite this case because it is one where explanation by memes would look so easy. We need only posit a new meme which successfully invades a population that has no immunity to it, a meme which then declines later as that immunity develops. The meme's success is due to its own strategy – presumably produced by a mutation – not to any fact about the people it infects. We need not look at these people, except perhaps to assess the general strength of their immune systems. We need not relate this meme to any other cultural viruses or parasites currently infesting this population nor to that population's earlier or later cultural history. We certainly don't need to reflect on human psychology generally, still less to look into our own hearts to see what we might learn about such conduct. We simply place the whole causation outside human choice.

But, placed there, the meme story simply gives us no explanation at all. What we actually need, when we are trying to understand such a case, is to grasp how people could begin to think and act in this way in spite of the beliefs, customs, laws and ideals which had stopped them doing so earlier. We need to grasp the psychology of persecution from (so to speak) the inside, to get some idea what it was like to contemplate such actions. And we need this, not just in relation to witch-hunting but for human conduct in other times and places too, not least in our own lives.

Understanding these activities does not mean discovering, by research, new facts about their causation by an imaginary alien life-form. No such outside entities are needed to explain human actions. *Understanding* those actions is centrally an exercise in self-knowledge, an exploration of what de Tocqueville called 'the habits of the heart'. Of course culture comes into the explanation, but we cannot save ourselves the trouble of this painful form of enquiry by invoking mythical culture-units. What goes on in these investigations – and what has in fact gone on over this topic of the witch-craze – is that historians look sympathetically for people's reasons, exploring the expression of them in the documents left by an earlier age. In this case they notice things like the various fears raised by an epoch of violent change, especially fear of the rising status of women at the time – the disintegration of earlier belief-systems – the horror caused by civil wars – the rising interest in the idea of a devil and also (of course) the various political interests that could be served by scapegoating a group without power.

All these explanations have some force, all of them do something to make the story more intelligible. None of them explains it completely, but of course this does not mean that historians have failed. Their partial explanations still throw a lot of light both on this particular phenomenon and on parallel cases such as anti-Semitism and xenophobia. Seriously worked out, they shed light on the general habits of the human heart. Whether we choose to call that light-shedding *scientific* is not, perhaps, particularly important.

CONCLUSION

There are two reasons why I have thought it worth while to examine this rather bizarre memetic programme at some length. First, in relation to the general theme of this book, it is an excellent illustration of the mess that tends to result when models drawn from the physical sciences are drafted in without good reason to explain human behaviour. Second, in relation to the current state of psychology, it seems important to emphasise that projects like this are not just idle but are a serious distraction from the sorts of explanation that we really need. And such distractions are particularly dangerous at a moment when psychology stands a chance at last of growing in more useful and realistic directions.

The lifting of the behaviourist taboo on serious discussion of our inner life has released a remarkable flood of interest, a fertile crop of new suggestions which surely needs to be encouraged. This change makes it possible to reverse the irrational distortion which took place a century back when behaviourism was allowed to drive out a wide variety of psychological enquiries ranging from those pursued by William James and John Dewey to the whole Freudian movement, on the quite mistaken ground that it was more 'scientific' than they were.

We will look more fully at the meaning of such claims and the deadening effect that they still have on controversy in Chapter 14. Meanwhile, the point that I have been making in the last two chapters is that the general tendency to use these claims as a form of quarantine, cutting off what is 'scientific' from its context in the rest of thought, cannot work. I have been pointing out the imaginative visions which underlie those sciences with a view to breaking the habit of polarisation that separates them radically from the humanities, producing an unwinnable 'war of the two cultures'. As I mentioned, this feud saddles the young with a painful and unnecessary dilemma, forcing them to specialise on one side or the other of the divide – a choice which they have to make much earlier and much more completely in English-speaking countries than they do in most European ones.

It should be obvious that, in trying to close this culture-gap, I am not attacking science but emphasising its importance in our culture. I am trying to explain why it is so influential and in what ways that influence has been working. The reason why I have drawn attention to the imaginative and metaphysical background of our scientific tradition is that this background deeply affects the rest of our thought, contributing strongly to the shape of our moral attitudes and our value-system. Science is not an isolated, autonomous, omnicompetent castle but an organic part of our total world-view. That is why we all need to be conscious of it. The visions that underly it ought to get far more attention than they now do in discussions both of literature and of the physical sciences themselves.

Part II

Mind and Body:
The End of Apartheid

7

PUTTING OUR SELVES TOGETHER AGAIN

THE SIZE OF THE PROBLEM

In our time, then, something called 'the problem of consciousness' is beginning to worry scholars in a number of disciplines, including those in which, until lately, that word was not supposed to be heard at all. Territorial disputes are even breaking out over whether this new problem is the property of scientists or philosophers.[1] There are in fact many problems involved, not just one. But the most interesting of them are more or less bound to have both scientific and philosophical aspects, so that specialists will have to try and co-operate over them, hard though that may be. And, in what looks like the central and most difficult puzzle of all, both these aspects are surely present.

That central puzzle is not about 'how consciousness evolved', nor is it about 'how we would know it was there if we didn't happen to be aware of it already', though both those questions have raised a lot of dust. The central worry is: 'How can we rationally speak of our inner experience at all? How can we regard our inner world – the world of our everyday experience – as somehow forming part of the larger, public world which is now described in terms that seem to leave no room for it? On what map can both these areas be shown and intelligibly related?'

This is a genuinely difficult issue, not a false alarm. It will not yield simply to familiar methods and a good injection of research money. But it is not desperate either. We can think about it if we are willing to stand back, to look at things from farther off and to admit the size of the question. Current panic about consciousness arises largely from trying to treat it as if it were a much smaller issue than it is.

The analogy that I would like to suggest here comes from geography. We are not looking for the relation between two places on the same map. We are trying to understand the relation between two maps of different kinds, which is a different sort of enterprise. At the beginning of an atlas, we usually find a number of maps of the world. Mine gives, for instance: world physiography (structure and seismology), world climatology (mean annual precipitation, climatic fronts and atmospheric pressure), world vegetation, world political, world energy, world food, world air routes and a good many more. If we want to understand how this bewildering range of maps works, we do not need to pick on one of them as 'fundamental'. We do not need to find a single atomic structure belonging to that one map and reduce all the other patterns to it. We do not, in fact, have to do once more the atomising work that has already been done by physics. Nor will that work help us here.

What we do need is something different. We have to relate all these patterns in a way which shows why all these various maps are needed, why they are not just contradicting one another, why they do not just represent different alternative worlds. To grasp this, we always draw back to consider a wider whole. We look at the general context of thought and life within which the different pictures arise. We have to see the different maps as answering different kinds of question, questions which arise from different angles in different contexts. But all these questions are still about a single world, a world so large that it can be rightly described in all these different ways and many more. It is that background – not a common atomic structure – which makes it possible to hold all the maps together. The plurality that results is still perfectly rational. It does not drop us into anarchy or chaos.

When we are using the atlas, we can make these connections quite easily so long as we remember this wider context, because the same coastlines appear on all of them. This helps us to relate the various pictures to the world they represent. If, however, we forgot that context and try to examine a smaller area on its own, we would be in trouble. Someone (for instance) who decides to investigate a particular square of Central Africa or Australia by cutting out and magnifying the parts of all these maps that showed that area, would find that the lines on many of these squares would not seem to bear any meaningful relation to one another at all. The political map, especially, might just show a single straight line running right across the whole area – perhaps pink on one side, blue on the other – something quite extraordinary which no other map ever shows. (There are no straight lines in nature.) And the airline map might be rather similar, but showing a different line.

At this point, enquirers might give up. They might say that the maps disagreed so badly that there could only be one of them that was correct and fundamental – only one map which really showed the world at all – namely, the one that they had backed in the first place. And this seems to be very much the way in which many people are now trying to investigate the problem of consciousness.

In considering that problem our project (as I have suggested) is not to relate two things which already appear on the same conceptual map. It's about how to relate two maps that answer questions arising from different angles. Consciousness is not just one object, nor one state or function of objects, among others in the world. It is not (as people often suggest) a function roughly parallel to digestion or perspiration. It is the condition of a subject, someone for whom all those objects are objects. The questions it raises are therefore primarily about the nature of a person as a whole, a person who is both subject and object.

When (then) we ask how consciousness can be a feature of the world, we are asking how we ourselves – as subjects – can be both items in the world and aware of it as a whole? This is not a factual question. It is a question about how to find convenient ways of thought, about how we can best think about an item that has this double position. How does that awareness fit – conceptually – both into the world and into the rest of our complex nature? Science itself has so far depended on and nurtured realism in the sense of a belief that the world is actually there. The problem now is, how to be realist about subjects – which are, it should be stressed, themselves natural phenomenona, not some kind of invented spooks.

Consciousness, then, is not just one more phenomenon. It is the scene of all phenomena, the place where appearances appear. It is the viewpoint from which all objects are seen as objects. The first set of questions that arise about it are questions about ourselves. These questions have to come before more strictly scientific questions about its place in the outer world, such as how it evolved. And it is important that this consciousness which we must look at contains a whole mass of more dynamic items besides simple appearances. It contains complex patterns such as emotions, efforts, conflicts, desires – aspects of our active participation in what goes on around us. If we once start to sort these things out, we shall probably have to think carefully about agency and free will as well as perception. And (if one may mention it) about time too.

PRIOR TABOOS AND THEIR DECAY

All this is very hard to deal with today. Current scientific concepts are not adapted to focusing on subjectivity. Indeed, many of them have been carefully adapted to exclude it, much like cameras with a colour filter. People have not, of course, actually supposed up to now that they were unconscious. The fathers of modern science took their consciousness for granted, since they were using it to practise science. Even the Behaviourists, who did try to deny this awkward factor, still tacitly presupposed it. Although they were officially epiphenomenalists who believed that their thoughts could not affect their behaviour, they still assumed that it was worth their own while to go on thinking and reading about scientific subjects, as if their thoughts might really have an influence on the world. They assumed that conscious beings were there to receive and understand their words. They assumed, too, that they had useful colleagues – that testimony from other conscious beings was a sound source of scientific knowledge. And so forth.

This background community of conscious subjects has, in fact, to be presupposed if any form of connected thought is to be possible at all. Until very lately, however, these subjects were not seen as something that science itself could study. They had been shut out of its domain, with good reason, during the Renaissance. Galileo and Descartes saw how badly the study of objects had been distorted by people who treated these objects as subjects, people who credited things like stones with human purpose and striving. So they ruled that physical science must be *objective*. And this quickly came to mean, not just that scientists must be fair, but that they should treat everything they studied only as a passive, insentient object.

We know that this abstraction made possible three centuries of tremendous scientific advance about physical objects. Today, however, this advance has itself led to a point where consciousness has again to be considered. Enquiries are running against the limits of this narrowed focus. In many areas, the advantages of ignoring ourselves have run out.

This has happened most notoriously in quantum mechanics, where physicists have begun to use the idea of an observer quite freely as a causal factor in the events they study. Whether or not this is the best way to interpret quantum phenomena, that development is bound to make people ask what sort of an entity an observer is, since Occam's Razor has so far failed to get rid of it. This disturbance, however, is only one symptom of a growing pressure on the supposedly subject-proof barrier, a pressure that is due to real growth in all the studies that lie close to it.

The pressure is naturally strongest in the social sciences, especially in psychology. As a direct consequence of the success of physical science, the social sciences were initially designed to imitate its methods as far as possible. Sometimes this has worked well, but not always. Social investigators who have tried to confine themselves to the methods used by their physical colleagues have repeatedly run into trouble at certain points because they found that they simply couldn't make progress without considering their subjects *as subjects* – their people as people. At these points, not even the most objective observer could dismiss the subjects' own point of view as irrelevant. It had to be acknowledged that people were in some ways different from stones.

Yet for a long time it seemed that the behaviourist project was the only possible way in which psychology could make good its status as a science. Its ideology was thus so fervently launched and so fiercely policed that its principles prevailed for the best part of a century, long outlasting the detailed work that was supposed to support them. During that time, it was as much as a social scientist's career was worth to be caught taking what B.F. Skinner so oddly called 'an anthropomorphic view of man'.[2] The methodological argument that underlay this view ran much like this:

Only what science studies is real,
Science cannot study consciousness,
So: Consciousness is not real.

And, though the conclusion has officially been abandoned, both these premises are still widely accepted. Hence much of our present difficulty. The church of academic orthodoxy now officially lets its members see these problems. But if we want to see them clearly, we are going to need a more suitable set of concepts. The terms which served seventeenth-century thinkers for dismissing conscious subjects from scientific attention are not likely to be the best ones for bringing them back into focus now.

THE RETURN OF THE FIRST PERSON

We need other ways of thinking. We have to stop thinking of consciousness as a peculiar, isolated feature of certain objects – as just one particular state or function of certain organisms – and start to think of it rather as a whole point of view, equal in size and importance to the objective point of view as a whole. And we shall not get far with this if we start our investigation by worrying about whether we can recognise consciousness in other people – about the so-called 'problem of other minds'.

To suppose that we have a problem about the existence of other minds is to be in trouble already because it is to have started in the wrong place – Descartes' wrong place. If we once sit down in that place, we shall never get rid of the problem. (Bertrand Russell, who was wedded to this starting-point, never did get rid of it.) This approach conceives of minds – or consciousness – unrealistically as self-contained, isolated both from each other and from the world around them. It is terminally solipsistic.

To avoid this unreal isolation, we had better attend in the first place to the examples that are most familiar to us, namely, our own experience and its relation with that of others familiar to us. Consciousness is not something rare and exotic found only in experimental subjects or in scientific observers. Nor does it only show us a few special phenomena such as colours and dreams and hallucinations. It is not primarily an observation-station. It is the crowded scene of our daily lives. And the main dramas going on in it do not concern just observation or per-ception but quite complex, dynamic currents of feeling and efforts to act. If we mean to do justice to this complexity, we have to take seriously the rich, well-organised language which we use about it every day. That language does not just express an amateur 'folk-psychology'. It is the indispensable working skeleton of all our thought – including, of course, our thought about science.

THE STARTING-POINT: SELVES IN SOLITARY CONFINEMENT

Perhaps this will become clearer if we look back again at the unreality of the traditional scheme which is now making this problem so hard. That scheme was, of course, Descartes' sharp division of mind from body. In order to make the natural world safe for physics, Descartes pushed consciousness right out of it into a separate spiritual world, treating each soul or mind as a spiritual substance, made of a stuff alien to other earthly items.[3]

Today, we cannot possibly put this extra entity, this dis-embodied mind back into the natural world, though Descartes' dualistic followers still try to. (The suspicion that we may have to restore it may account for some of the alarm that now surrounds this topic.) The trouble is not just that physics leaves no room for this kind of entity, but that thinking creatures could not possibly be isolated entities of this kind. Thought involves communica-tion. Cartesian egos, isolated inside their separate shells of alien matter, could never even have discovered each other's existence. *What thinks has to be the whole person, living in a public world.*

The 'problem of other minds' arises from positing this solipsistic self, which we might call Descartes' diamond. It is a hard, impenetrable but very precious isolated sentient substance which sits at its console in its windowless tower communicating with other, similarly secluded diamonds by signals run up between towers and relayed to these beings by a perpetual miracle. Real people, by contrast, are embodied beings living in a public world.

Descartes' reason for introducing this awkward kind of soul was his need to compromise between three conflicting kinds of demand – the position of traditional Christianity, the need to segregate physics from other studies, and the demands of everyday language. The diamond never fitted any of these systems at all well and as time has gone on the misfit has grown steadily worse. It was always an unstable notion. When it was taken seriously, it has usually tended to expand into absolute idealism – the idea that spiritual diamond-stuff was actually the stuff of the whole universe, a stuff that underlay physical matter as well as souls. Hence the mentalist tradition that runs through Leibniz, Berkeley, Hume and Hegel to modern phenomenalism.

Today, by contrast, many educated people, and in particular many scientists, assume that the opposite metaphysic – materialism – has simply conquered this whole trend and now reigns unchallenged. But it seems increasingly clear that one extreme view is no more workable than the other. Materialism and idealism are equally the products of dualism. That is why the unstable notion of an isolated self has not gone away. The ghost still haunts the machine because the machine has not changed thoroughly enough to do without it.

What this all shows is that the matter is complicated. The very word 'self' is of course not an everyday word but one coined by philosophers who were trying to make certain difficult questions statable. The trouble is that these questions are rather varied. For some purposes, such as 'self-defence', the self clearly includes a body. For others, such as 'self-knowledge', it seems not to – and so forth. Theorists – such as Descartes – have been tempted to try and find a standard entity for it to refer to. They have hoped for a Procrustean bed guaranteed to fit all these uses. And in modern times the hope has been that something narrower and simpler still would emerge – a concept modelled on the most abstract scientific terms such as the names of physical particles. This is surely a mistake because these scientific terms are designed only for a specially abstract context. Even in biology it has not proved possible to standardise terms like 'species' in this way because they refer to things in the world which have to be looked at in a variety of lights, and this is still more true of notions that play an important part in the jungle of human life. Attempts to impose simple, arbitrary definitions on the subtle and versatile words that play an important part there – words such as *person, reason* and *feeling* – are inevitably futile.

What is needed in such cases is rather to make clear just what the context of the particular enquiry is – why this particular question is arising. Richard Sorabji, noting the bewildering variety of views about soul and body held by various Greek philosophers, and sometimes by the same philosopher on different occasions, comments:

> I believe that this is perfectly natural because of the reflexive things that people want to say about the self. The self can be attached to the self, or conscious of the self (Hierocles), it can construct a self (Epictetus, cf. Plutarch) or direct a self, even directing it to another self (Plotinus). It is not always the same kind of self that is engaged at either end of these reflexive relations.
>
> ('Soul and Self in Ancient Philosophy' in *From Soul to Self*, ed. M. James and C. Crabbe, London, Routledge, 1999, p. 15)

In short, we always have a rather limited vocabulary here for a complicated topic. We can, however, make it work if we attend to the complications and make it clear on each occasion what we are trying to do with our tools. I have tried to do this about awkward words such as *self* in this book, though I am sure I have not always succeeded.

8

LIVING IN THE WORLD

WHY THE GHOSTS?

Solitary pseudo-souls have been a surprisingly persistent image, often welcome to the individualism of our age. These chips off Descartes' diamond can be found sticking in the works of many supposedly secular conceptual systems. For instance, there is existentialist ethics, where the will, which is supposed to be solitary and independent of the entire natural constitution, is alone considered to be the authentic person.[1] Such doctrines disconnect the conscious self entirely from the world around it. There are also others, popular today, which do allow for a social context but deny a bodily one. In sociology for instance, there is still quite a widespread belief that human behaviour can only have social causes, not biological ones, so that the constitution of people's bodies can have no effect on their personality. And again, there is post-structuralist literary criticism, in which the arbitrary relation between sign and signified is supposed to cut literature off from the everyday world which it might seem to be about, leaving it in some kind of spiritual mid-air as a purely cerebral network of texts. From the materialistic side, too, there is the medical model that represents patients simply as bodies sent to be chemically and physically repaired, on the assumption that their conscious life is a separate matter which won't be relevant to that process.

Most remarkable of all, however, are the electronic dreams of down-loading human consciousness onto computers, dreams which take it for granted that the personality is a kind of software that does not need the body because it can be run with equal ease on any kind of hardware. It may then even be sent ambitiously off to outer space, there to achieve eternal enlightenment, freed from the limitations of the earth.[2] But why – we might ask – should earthly creatures go so far in order to do that?

> Does the fish soar to find the ocean,
>> The eagle plunge to find the air –
> That we ask of the stars in motion
>> If they have rumour of thee there?
>
> Not where the wheeling systems darken,
>> And our benumbed conceiving soars! –
> The drift of pinions, would we hearken,
>> Beats at our own clay-shuttered doors.
>
> The angels keep their ancient places: –
>> Turn but a stone and start a wing!
> 'Tis ye, 'tis your estranged faces.
>> That miss the many splendoured thing.
>>> (Francis Thompson, 'The Kingdom Of God')

I cannot spend more time here on these doctrines, some of which I have discussed elsewhere.[3] It seems better now to do the more positive thing which is always needed when one attacks a dominant image and to try, however crudely, to suggest how we might think instead.

What kind of image might we use, then, as a corrective for this strange isolationism? Well, we could do worse than listen for a start to A.E. Housman's view of the matter:

> From far, from eve and morning
>   And yon twelve-winded sky,
> The stuff of life to knit me
>   Blew hither: here am I.
>
> Now – for a while I tarry
>   Nor yet disperse apart –
> Take my hand, quick, and tell me
>   What have you in your heart?
>
> Speak now, and I will answer;
>   How can I help you, say;
> Ere to the wind's twelve quarters
>   I take my endless way.
>
> (A.E. Housman, *A Shropshire Lad*, xxxii)

We do not need to take up Housman's whole world-view in order to grasp his point. There may well be parts of that we do not want at all. But this poem surely gets two striking features right:

1   This 'I', this subject of Housman's, is as far as possible from being a pure, isolated diamond. It is a composite being. It has been formed by and out of the natural world. It is still a part of that world and wholly dependent on it.
2   It is continuous with that outside world in being essentially social – not solitary. This self never started life alone. It does not have any problem of other minds. It asks someone flatly (as any of us might) 'what's on your mind? what

have you in your heart?'. It speaks directly to another, with whom it shares a language and for whom it is deeply concerned. It is, in fact, a genuine member of our species.

THE IMPORTANCE OF BABIES

At this point, we might do well to remember that this is a species whose members, as babies, communicate with other people long before they try to handle inanimate objects. They also learn other people's names before they learn their own names. And they learn to talk about other people's mental states long before they become introspective enough to discuss their own. From birth, they are equipped – just as other young social animals are – with the right capacities for expressive behaviour, and with the power to interpret that behaviour in others. These capacities are not imitated from others later. Babies born blind smile, and non-blind babies can at once interpret smiles. Deaf babies cry. And so forth.[4] This innate repertoire makes it possible for human babies, just like other primate babies, to start communicating as soon as they begin to be aware of the world at all and long before they take any other sort of action.

That is how human infants manage to take in, quite directly, the mass of facts about other people's attitudes which will be the foundation of all their social knowledge and which will – among other important things – also make it possible for them to learn to talk. They do not need inference to do this. In fact, in our species, any social inference there may be is primarily directed inwards, from our knowledge of others to ourselves, rather than outwards by analogy from ourselves to them. We gradually learn to apply to ourselves the words which we already use to describe other people's moods and characters.

We can, indeed, then have a problem of self-knowledge – a 'problem of one's own mind'. We are often amazingly ignorant about our own condition and our own motives. This ignorance can become important later in life and when we are more mature it is our business to deal with it. But there is no original problem of other people's minds. For us humans at least (whatever may happen elsewhere in the universe), the whole use of language depends on, and arises out of, the deep, innate, unshiftable sense that each of us is only one among others who experience the world in roughly the same sort of way that we do ourselves. This sense is constitutive of our thinking, not a dispensable part of its content. Autistic people, who apparently lack that sense, tend to have great difficulty in using language at all.

DO WE NEED PROOF THAT WE ARE NOT ALONE?

Do we need some sort of proof for very general assumptions such as the assumption that we live in this public world? If so, one might ask, for a start, whether it is this assumption that needs proof or its opposite? If someone decides to assume that he does indeed live alone and has invented the companions that he has so far believed to surround him, would that assumption be less in need of defence than assuming a shared world or more so? Would it somehow be more economical?

We certainly do not know enough about the initial conditions to say which of these situations would be abstractly the more probable in an imaginary world. But in considering the world that we do know we are better off. The way to test such assumptions is to ask which of them makes more sense in the context of that world. The only kind of proof that they can have consists in showing that they are necessary in order to make thought and language possible.

We can see this well over two familiar assumptions:

1   That nature is regular or lawful – that unobserved facts will go on being like the observed ones so that the future will be more or less like the past, and:

2   That in general we can trust our faculties – that our senses, memory and reasoning powers are not misleading us radically all the time.

Of course sceptics are right to say that propositions like these are too wide to be checked in experience. No check could conceivably be adequate. They sometimes conclude that it is irrational to accept these assumptions. Indeed, they sometimes hint that if we were less lazy we would not believe them. Thus Hume, rejecting the reality of causation, concluded that 'if we believe that fire warms, or water refreshes, it is only because it costs us too much pains to think otherwise'.[5]

But this sets an unreal standard of rationality. The objection to dropping these basic assumptions is not just that thought becomes hard without them but that it stops altogether. Hume makes it sound as if we could go on on this path if we tried, just as we could, with an effort, dismiss a particular belief which we found to be prejudiced and unfounded. That kind of limited dismissal, however, leaves the whole background of our ordinary beliefs still standing. It still leaves us a world. By contrast, the notion of dismissing that whole background – of losing the basic conditions that make any experience reliable at all – produces a total conceptual vacuum in which the dismissal itself would lose all meaning along with everything else. There is no conceivable point to which thought would then move. Attempts at disbelief of this kind would not just run into emotional difficulties due to laziness. They would hit a logical block, like attempts to square the circle.

Rationality cannot require this. It demands that we should accept the conditions which are evidently necessary for reasoning, not that we should reject them in a desperate attempt to get an irrelevant sort of proof. Indeed the word proof itself simply means 'test', and different kinds of test are needed for propositions that do different kinds of work.

How does this general point about what is rational and what is not – what needs empirical proof and what does not – bear on our present argument? The fact that each of us is not alone in the world – that we live among others sufficiently like us to communicate with us – is one more of these basic conditions which are needed to make human thought possible at all. It is unlucky that Descartes failed to see this when he shaped his systematic doubt and thus came to forge the unreal, solipsistic conception of the mind that we have been discussing.

Descartes' oversight is remarkable in view of the stress he laid on the fact that thought needs language. For language is clearly a corporate, social phenomenon. Anyone who is in a position to say a sentence such as, 'I think, therefore I am' has to be heir to a rich and widely-shared linguistic tradition and must therefore be a member of a widespread company of similar beings. More generally still, all intelligent animals are social animals. Their thought is always a set of tools forged by a whole community. But for people in particular, the elaboration of these tools – the richness of a language that has required much time and trouble to evolve – makes the idea of a solipsistic life empty and inconceivable.

AGENCY, FATALISM AND DIGNITY

These two propositions – physical and social continuity with the surrounding world – are not really contentious. They are obvious and fundamental truths about human subjects. In practice we all accept them. But many theorists have resisted admitting them, because they felt that this biological and social dependence on the outside world weakened the subject's dignity intolerably. And it is quite true that our dependence on the world does mean that we cannot have the guaranteed metaphysical autonomy which Descartes' diamond was supposed to enjoy.

In spite of that dependence, however, we see that Housman's 'I' also still manages to function vigorously as an active agent. Though this 'I' knows that it is composite and transient, a weak, shaky fallible thing at the mercy of the natural world, it has no hesitation in offering to act. And there is no reason why it should have any such hesitation. It asks 'what can I do?', and this is a real question. It means to try and to do it.

This self is not held back from making its offer by any theoretical fatalism, any conviction that it is really just a helpless, anonymous cog in the cosmic machine. (Housman, like the Stoics, was fatalistic about outside conditions, but never about internal freedom.) And as we all know, there is absolutely no reason why it should be held back by that kind of fatalism. Someone who is faced by a friend needing help is not likely to reply that, since we live in a deterministic world, he can't actually do anything because he is unfortunately unable to act freely. And if anybody did make that reply, the right name for it would not be 'scientific correctness' but 'humbug'.

TWO SIDES TO OUR HEAD

This two-sidedness of our nature certainly is very surprising. How are we to remember that we are, on the one hand, earthly organisms, animals like others operating within a physical pattern, a pattern which we can treat as fixed and predestined when we take the outside viewpoint – but that we are also, on the other hand, *agents*, active beings who not only can but must choose what to do? Standing at our inner viewpoint we must continually take decisions. This is the freedom of practical thinking, a freedom which we need, not only in order to act, but in order to think as well. These practical choices cannot possibly be reduced (as Skinner sometimes tried to reduce them) to simply predicting what we might be expected to think or to do.[6] That is something quite different.

It is no wonder that we have evolved two rather distinct ways of talking for these two different contexts or that, much of the time, we use these two languages side by side without noticing any clash between them. As Kant said, we use the deterministic one most in speculative thinking – especially when we think we are doing something called 'science' – and the active one most for practical life. Often we can split ourselves neatly in two so as to duck possible contradictions between them. But we cannot do this all the time. And if we are interested in the larger scene – if we want to put the whole jigsaw together – we cannot avoid the problem of how to relate these two modes. Above all, we have to relate them when we think about personal identity – about what is, and what is not, essential in our lives, about the kind of being that each of us is as a whole.

HORNS OF A DILEMMA?

In recent times, however, scholars have not tried very hard to take that comprehensive viewpoint because they have preferred to make war. This war has usually been seen as being about which stuff the world is made of – rather on the pre-Socratic model. It was assumed to be made of one kind of stuff – either of mental diamond-stuff or of matter. Materialists and idealists shot arguments at each other to decide which stuff should prevail and which language could be hailed as describing the real world. The other language could then be sidelined as just something handy for ignorant people but mildly misleading – in fact, a folk-language. It might then still be used in everyday life until a better code was invented but we were urged to remember not to take its concepts too seriously. As Bishop Berkeley kindly put it, we could 'think with the learned and speak with the vulgar'.[7] And at present 'thinking with the learned' is expected to mean thinking as a devout materialist, exclusively in terms of insentient objects.

The trouble about this way of splitting our lives is that the vulgar, the folk, are *ourselves*, not poor relations whom our learned selves can disown. They constitute the core of our being. The learned ways of thinking and talking which we develop for certain limited purposes are secondary extras. These techniques cannot displace our basic structure of agency and responsibility because scholarly activity itself relies on the use of that structure, just as much as other kinds of activity do. The practice of science is itself a *practice*. It has to be understood as a way of acting freely, deliberately and responsibly. Otherwise it cannot be understood at all. Our speech is not alone in having to follow the vulgar, everyday pattern here. The central structure of our thought does so as well.

It is interesting how profoundly the secondary, scholarly ways of thinking can vary from age to age. Berkeley disowned the vulgar from a position very different from today's. Berkeley was an idealist who believed that the language of spirit had won the battle. To support its triumph, Berkeley gave good reasons why the materialist position couldn't be right, reasons why you cannot actually have a world of objects without subjects. Hume followed him by showing that it is actually quite hard to prove that the material world exists at all, if you once start to doubt it, and that in many ways it may be more economical to do without it.[8] In philosophy, in fact, the idealist army won some well-deserved successes up to the end of the nineteenth century.

Later, however, the fortunes of war went against it. Materialist troops seized the disputed territory and occupied it for most of the twentieth century, laying down assumptions that are still today the official creed of the Church Scientific (as T.H. Huxley called it). Yet the reason why we are worrying about consciousness today is that this occupation hasn't worked very well either. The occupying troops themselves – the scholars working in areas where questions about conscious subjects keep coming up – are not happy. They report that the natives are restless. To drop the metaphor – the simplifications which materialism introduced, and which often conferred great advantages at first, have in many areas been exploited to their limit. Their attendant disadvantages are now coming forward. The approach must somehow be changed.

HAVING IT BOTH WAYS

How do we change it? A favourite suggestion for this at present is to bring out yet a third stuff, called information, making it the successor to both mind and matter. There are two reasons why this doesn't work. First, it only looks plausible because the word *information* is so wildly ambiguous that, when it's sloppily used, it can look like a stand-in for either of its predecessors.[9] And second, it misses the point that this whole rivalry was misconceived in the first place.

There is no competition. There does not have to be a single basic stuff. Materialism and idealism are not the names of warring tribes but the names of half-truths, neither of which can really be swelled into a comprehensive system. The real question is not 'what stuff is the world made of?' but – as Tom Nagel says – 'how to combine the perspective of a particular person inside the world with an objective view of the same world, the person and his viewpoint included'.[10] How can we bring these together so as to give a reasonably intelligible whole instead of leaving them loose so that they slide around and bang into one another?

Of course it was natural to hope that one single conceptual scheme would provide a comprehensive, all-purpose explanation. This would have satisfied the demand for simplicity which is one central aim of our reasoning, and which many people still seem to see as the only one. Seventeenth-century thinkers were convinced that, at the deepest level, the world was, as a matter of fact, simple in this way – a faith which seems still to have appealed to Einstein. And of course they followed that hope to tremendous effect. But there could be no guarantee that their conviction was right.

Today, many physicists see much less reason to make this wholesale assumption at all. For all we know (they say), at the deepest level the world may go on for ever becoming more and more complex. The great reason for simplifying it is not that we have any advance guarantee of results. It is heuristic. It is just that, when simplifying does work, it greatly advances our understanding. It is therefore always worth while looking for simplicity. But that is very different from knowing that it is there to be found.

Many people today still believe that rationality offers us only the two competing schemes of dogmatic idealism and dogmatic materialism. They assume, not just that the universe has to have a single ultimate structure, but that that structure must be one of these two that we have already started to use. When we consider how continually schemes of thought have changed in the past and how much effort still goes into reshaping them this does seem an extraordinarily bold belief. Rationality does not always demand simplification, and it certainly does not always demand elegance. Rationality is concerned with finding a sane balance between various sorts of considerations, with describing the world realistically and with adjusting means to ends. In science, and in enquiry generally, this adjustment means finding a conceptual scheme that works, one that suits your subject-matter. Occam's Razor cannot be the only tool on our bench. As Aristotle rightly said, it is the mark of an educated man to look for just so much exactness in every enquiry as the topic that you are enquiring about admits of – not asking a literary critic for geometrical proof or vice versa.[11]

CATEGORY TROUBLE

In general, no doubt, we all admit this need to recognise complexity. But there is one particular kind of complexity which is quite hard to fit into current notions of what is scientific. This is complexity that involves more than one conceptual scheme and, still more, complexity involving concepts in different categories.

That is the kind of difficulty that I tried to indicate in my opening example of the different kinds of map. Somebody who tries to account for the lines on the political map in physiographic terms will not succeed, and will not do any better if they try to reduce physiographic language to something more basic still such as particle physics. Very simple maps are certainly available and they do have their attractions, as was pointed out in *The Hunting of the Snark*:

> 'What's the use of Mercator's North Poles and Equators,
>   Tropics, Zones and Meridian Lines?'
> So the captain would cry: and the crew would reply,
>   'They are merely conventional signs!'
>
> 'Other maps are such shapes, with their islands and capes!
>   But we've got our brave captain to thank'
> (So the crew would protest) 'that he's bought us the best,
>   A perfect and absolute blank!'.[12]

But the uses of such maps are limited. For instance, if we want to know the explanation of that mysterious straight line on the political map – which in itself appears so beautifully simple – the only place where we can find it is in the history of a certain treaty and in the complex colonial system that lies behind that treaty. Here, explanation has to move outward to that much wider and more subtle background, not inward to the atomic structure. Treaties cannot be explained in terms of contours or vegetation or electrons, nor of the neurones of the people who make them, any more than vice versa. They are not just a kind of casual folk-shorthand for those things either. What is needed at this point is systematic talk about human history and human purposes. And this is talk that proceeds from a quite different angle.

This discontinuity between different viewpoints and different languages is not something imposed by philosophers. It is not an alien trouble brought into science from the humanities. It is an unavoidable difficulty that is well known to arise whenever people from several different disciplines have to work together on a common problem. Modern specialisation has greatly increased it by multiplying and narrowing the disciplines. It can arise, too, within a single enquiry, about the ambiguity of some abstract term such as time, space, evolution, infinity, particle, wave or function. Terms like these have various possible meanings, arising from different contexts of use, meanings which have to be disentangled when their role in a particular science begins to cause trouble.

Our difficulties about consciousness are simply more extreme examples of category trouble than most of these. This gap between the concepts that are used mainly from the subjective viewpoint – including those that deal with action – and those used from the objective one is indeed probably the most puzzling conceptual gear-change that we ever have to deal with. But even the minor clashes can be quite awkward, because their conceptual background always needs attention as well as their factual foreground. And sorting out conceptual tangles cannot help being philosophical business.

That does not mean, of course, that it can only be done by philosophers. Scientists often do it very well. In fact, the greatness of great scientists has often centred on just this skill in clarifying crucial concepts, especially in resolving category-problems that arise about them. Until lately, these thoughtful scientists were quite content to say that they were doing philosophy. They considered this as one of the things that they normally had to do. They knew when they were using philosophical tools. People like Faraday and Maxwell, Bohr and Einstein, Heisenberg, Darwin, T.H. Huxley and J.B.S. Haldane understood that every branch of science has its own philosophical problems. They knew that these sciences very often need help from methods other than their own in resolving them, and they readily acknowledged that outside help.

The conclusion surely is that problems are not private property. They belong to anybody who can help to solve them. The advance of specialisation makes it harder to grasp this today, but it certainly does not make it less necessary. The fact that conscious beings are found in the world is a natural fact like any other. We have to accommodate it somehow. While we ignore it, we are not just left with a divided world but also with a disastrously divided notion of ourselves.

9

THE STRANGE PERSISTENCE OF FATALISM

WHAT KIND OF FREEDOM?

This division within our notion of ourselves is notoriously most acute over the 'problem of free will'. That problem, as we now conceive it, largely arises out of our current habit of splitting human beings into separate mind and body and then forgetting about the mind. This makes it appear as if the body goes on on its own unaffected by our thoughts, which are merely a side-effect – a view which is called epiphenomenalism.

Epiphenomenalism is not a sensible idea. It is incompatible with any real belief in our evolutionary continuity with the rest of nature. As we shall see, if seriously held, it would require extreme fatalism, invalidating not only most of our everyday thinking but also the practice of science itself. We need somehow to consider mind and body, not as distinct items but as complementary aspects of the whole person. (This conceptual shift is the real core of the current 'problem of consciousness'.)

What blocks this project is still the seventeenth-century reductive habit of trying to arrange explanations in a linear pattern to produce a complete and final 'fundamental' explanation – one which is now assumed to be in some sense physical. Complete explanations, however, are not available. Instead, what we always have is a set of partial and overlapping ones, dealing with different aspects, which can be adequately ordered and related. As we have seen, those partial explanations are not rivals, any more than the different maps of the world (physical, political, etc.) which we find in our atlases are rivals. We can combine these different accounts in the context of experience to make a usable whole.

DUALISTIC DIFFICULTIES

If we were considering a Chinese vase, we would distinguish many different aspects of it – its shape, its size, its colour, its history, its function and the material it is made of, perhaps also its beauty or ugliness. We would talk about these aspects in different ways. But this would not make us say that they are themselves separate items. Still less would we conclude that these items are so alien to one another that it is a painful and insoluble metaphysical mystery how they can interact at all. We would not be constantly worrying about the shape-size problem or the history-function problem. Yet when we think about those two aspects of a human being which we now call mind and body, our current tradition does impose this strange habit. The radical division that Descartes established between these two aspects is still entrenched in our thought.

At present, Descartes' division is, of course, being officially denounced on all sides. Yet, within the new ways of thinking that are being proposed, this split constantly reopens at a slightly different point. It does so because the reformers are still thinking in terms of distinct entities rather than of complementary aspects. Their schemes absorb the formal attributes of thought quite easily into the province of 'physical science' – though the word *physical* becomes somewhat strained in the process. But they still have a problem over something extra besides these attributes – something variously described as experience, subjectivity, phenomenal consciousness, qualia or consciousness itself – which will not go in the box. Just like Descartes, theorists still treat this remainder as an extra entity. Thus, for instance, David Chalmers, in an otherwise excellent article which makes beautifully clear the need to discuss subjective matters in terms that are not those of physical science, speaks of consciousness as 'the extra ingredient' and proposes a new place for it in ontological terms:

> Although a remarkable number of phenomena have turned out to be explicable in terms of entities simpler than themselves, this is not universal. In physics, it occasionally happens that an entity has to be taken as fundamental. [Thus] Maxwell and others introduced electromagnetic charge and electromagnetic forces as new fundamental components of a physical theory.
> (David Chalmers, 'Facing up to the Problem of Consciousness' in the *Journal Of Consciousness Studies*, vol. 2, issue 3, 1995, pp. 207 and 209)

Though the word *entity* is not a perfectly clear one, this comparison cannot help suggesting that what is to be introduced is not a distinct point of view on the whole of life – which is what is needed – but exactly the kind of quasi-material item, illicitly smuggled into the physical world, which philosophers are accustomed to pursue so sharply with Occam's Razor. And since modern theorists leave much less space for such an item than Descartes did, they are forced to accommodate it, as Chalmers then does, somewhere among the ambiguities of the weasel word *information*, in an awkward annex screwed on to science. *Information* looks as if it could cross the divide because, like *determine*, it has meanings in both the subjective and the objective realm. (We will discuss the word *determine* shortly.) But these meanings are in fact quite separate, so no bridge is provided.

All this worry is not really necessary. The present excitement about it is much as if our expert in Chinese vases were to believe that the function – or perhaps the history – of his vase must either be described somehow in quantitative terms, like its size, or be dismissed as a mere superstition. (Indeed the notion of function itself often does raise this kind of difficulty.) The trouble is that terms such as *consciousness* and *experience* belong in a different category to those of physical science, a category which is deliberately kept out of science just because, in terms of our whole thought, it has a more general application. Words like these become appropriate when we have moved right away from the specialised scientific angle to another wider and more basic one, namely, the angle from which we normally look at the world when we remember clearly – instead of deliberately forgetting – that we, and many others, are sentient beings within it.

Of course this wider point of view is not hostile to science. It is the background framework within which the sciences arise. The scientific point of view is itself an abstraction from it. The scientific angle is the one from which we attend only to certain carefully selected abstractions which are meant to be the same for all observers. When we move away from that specialised angle to the wider, everyday point of view we are not 'being subjective' in the sense of being partial. Instead we are being objective – i.e. realistic – about subjectivity, about the hard fact that we are sentient beings, for whom sentience is a central factor in the world and sets most of the problems that we have to deal with.

We often have reason to move away from this comprehensive everyday viewpoint in the opposite direction – the subjective one – so as to examine more closely our own and other people's experience. But none of the positions we can take is exhaustive. We always need to consider many of them together. In fact, human life is rather like an enormous, ill-lit aquarium which we never see fully from above, but only through various small windows unevenly distributed around it. Scientific windows – like historical ones – are just one important set among these. Fish and other strange creatures constantly swim away from particular windows into areas where we cannot see them, reappearing in other places where different lighting can make them hard to recognise. Long experience, along with constant dashing about between windows, does give us a good deal of skill in tracking them. But if we refuse to put together the data from the different windows, then we can be in real trouble.

FREE FROM WHAT?

That refusal is the root of our trouble on the topic that we call free will. This name is a bad one because the mention of freedom already gives the wrong picture. Freedom is a relation between two separate beings, an escaper and a potential controller. We can be free from prisons, or tyrants, or even diseases, because these are entities distinct from ourselves. But we cannot be controlled by our own bodies unless those bodies are, in just as full a sense, distinct from what we call 'ourselves'.

To speak of ourselves as controlled by our bodies is to accept a special notion of our selves which excludes those bodies. The words *free will* originally meant simply freedom from compulsion by other people, as in 'acting of one's own free will'. But they are now used quite differently, to narrow our notion of our self by excluding our own bodies from it. Freedom does not here have the modest meaning it has when we speak of being free from, say, an obsession, a compulsion or a prejudice. There we quite properly use the language of freedom, not just to reject these things but to redefine our true self as something distinct from them. But the language of free will extends this kind of rejection far more drastically to something which it would be much odder to exclude from our personal identity – namely, to our whole physical aspect. And it becomes particularly strange to reject this aspect if we are specially interested in our place in nature. Indeed, from the outset dualism has vitiated our understanding of evolution.

THE SLIDE INTO FATALISM

Descartes designed his notion of the self or mind as a distinct entity so as to account for the fact that we have to talk in different ways when we are describing these two aspects of ourselves. Because the two languages involved were already diverging strongly in his day, he divided these aspects by the extreme measure of calling them separate substances. But this language affects our whole conception of personal identity in a way which goes far beyond what he needed. This strong, simple dualism persists today even though in theory it is often rejected. Indeed it is currently being reinforced by computer imagery. Many people find it perfectly natural to say – apparently as a fact and not just as a metaphor – that mind is divided from body exactly as software is divided from hardware. This gives traditional dualism an even stronger grip, which sharply needs our attention.

On Descartes' model, the world that includes our bodies is one vast physical system, essentially a machine. And by calling it *physical* he meant, quite literally, just what is described in the science of physics. This is a far narrower sense of the word than the casual sense in which we speak of bodily pains as physical, or call an apple a physical object. Both pains and apples are foreign to the language of physics, which deals in terms such as equations describing systems of fields and forces. For Descartes, minds existed outside that system as conscious substances of a separate, parallel kind. But increasingly, later thinkers found these displaced entities embarrassing and shut them out from scientific talk. Thus, the imaginative picture which has shaped our supposed modern problem of free will shows human life, no longer as a drama where active people struggle against difficulties, but as one where they do not exist as distinct entities at all, only as areas of matter which are passive cogs, parts of a vast alien machine.

That is the picture which Richard Dawkins presents so forcefully at the outset of his book *The Selfish Gene*, writing that 'We are survival machines – robot vehicles blindly programmed to preserve the selfish molecules known as genes' (p. 10). Dawkins evidently does not regard this phrase as rhetoric but in some sense as literal fact, for he adds 'This is a truth which still fills me with astonishment'.

This kind of image, however, is not one that could be literally believed in. It belongs essentially to third-person talk. It is a way of thinking devised for describing other people. There is no way in which we ourselves could set about living if we really envisaged ourselves as cogs or vehicles. For people who are not actually paralysed, this pattern is too fatalistic to provide any usable view of life.

Fatalism is not just a belief in causality or an expectation of disaster. It is essentially a view about *effort*. Fatalism says that our efforts must always be useless because a power outside ourselves controls our destiny and will override all our attempts to act. In so far as it is serious, fatalism is meant to stop us bothering to make those efforts. And of course there are particular predicaments, such as paralysis, which do call for cessation of effort because they make certain actions genuinely impossible. But this pattern cannot be extended across the whole field of activity because our whole conscious life involves effort. Even voluntary submission requires effort, and so, of course, does suicide. Sentient life is essentially active. That, indeed, is why it has to be conscious in the first place.

This link between consciousness, effort and thought is a key point that has been badly overlooked in the flood of recent literature on the subject. Although some of our complex actions (say, in driving or playing the piano) are indeed more or less unconscious in the sense of being inattentive, the effort by which we devise and choose those particular actions – rather than others – in the first place is not. These actions would not be what they are if we, and many others before us, had not consciously and deliberately chosen to pursue certain particular aims rather than others. Though these movements can now be performed absent-mindedly, the shape that they take results from conscious responses in the past to challenges that were consciously recognised. Pianos (for instance) do not put themselves together and begin to play while everybody around them is asleep or engaged on something quite different. Nor does the fact that machines can sometimes be made to imitate these performances eliminate the link between conscious effort and thought, because a great deal of effort is necessary to enable the machines to do so in the first place. Quite as much conscious effort goes into chess-programs as goes into the play of their human opponents. Indeed, as we know, a computer is in general a device which can be made, by three month's hard work, to do in five minutes something which otherwise would have taken its operators a week.

That is why, as E.J. Lowe points out, it is disastrous to treat the existence of consciousness itself as an isolated, mysterious problem – as what David Chalmers identifies as the 'hard problem':

> Because Chalmers misconstrues what he sees as being the 'easy' problems of consciousness, he also misrepresents what he calls the 'hard' problem. According to Chalmers the 'hard' problem is this: 'Why doesn't all this information-processing go on 'in the dark', free of any inner feel?' (p. 203) Believing as he does that human thought and cognition in general are just a

matter of 'information-processing' of a sort which could in principle go on in a mindless computer, he is left with the idea that all that is really distinctive in consciousness is its qualitative or phenomenal aspect . . . And then it begins to look like a strange mystery or quirk of evolution that creatures like us should possess this sort of consciousness *in addition to* all our capacities for thought and understanding . . . *The sort of phenomenal consciousness which we humans enjoy but which computers and trees do not, far from being an epiphenomenon of information-processing in our brains, is an absolutely indispensable element in our cognitive makeup.*[1]

10

CHESSBOARDS AND PRESIDENTS OF THE IMMORTALS

THE ROOTS OF FATALISM

Mechanistic fatalism, which claims that conscious effort does not affect the world, is not, then, a realistic world-picture and it would probably not have been accepted if it had been taken literally and judged on its own merits. The reason why it has caught on so easily is that it fits the mass of rhetoric inherited from earlier, more limited fatalistic ideas – ideas that are powerful emotionally, but less ambitious intellectually than the current one. They do not claim to state universal scientific truths about the laws of nature. They simply display strongly a scene – which is recognised as poetic image rather than literal fact – in which human beings are helpless victims of a remote or malignant Fate-figure. Thus, as the Persian poet Omar Kháyyám put it:

'Tis all a chequer-board of nights and days
  Where Destiny with Men for Pieces plays:
Hither and thither moves, and mates, and slays,
  And one by one back in the closet lays.

(*The Rubáiyyát of Omar Kháyyám*,
trans. Edward Fitzgerald, first edition, stanza xlix)

This imagery is not specially religious. Its main point is not to name a god but to provide a kind of universal excuse, a chance to blame fate and play the role of victim. Even declared atheists often find this handy. Such talk suited well with Renaissance astrology, a world-view which was not really theistic at all. Thus, in John Webster's play *The Duchess of Malfi*, a character who has, in fact, been largely responsible for his own ruin complains that:

We are merely the stars' tennis balls, struck and bandied
Which way please them.

(Act V, scene iv, line 52)

We might have hoped that Descartes' attempt to sterilise the physical world by calling it a machine would have discouraged this superstitious way of talking. Yet, only a century ago, Thomas Hardy, in whose thought atheism was an explicit central theme, still ended *Tess of the d'Urbervilles* with a kind of affectation of belief, writing that 'The President of the Immortals . . . had ended his sport with Tess'.

Such personification comes very naturally to us because we are animistic. Our imagination does not necessarily see machines as impersonal. Real machines are, after all, always instruments of human intentions. Thus, though the machine image officially represents an impersonal order of fields and forces, it does not stop this personal way of thinking. Indeed, the cosmic opponent who is seen as somehow running the machine may loom even larger than he used to because he now carries all the impressive baggage of modern science. His realm now extends outward to the furthest limits of space-time and inward to the recesses of our nervous system. The mechanism revealed is so all-pervading that it becomes absurd to think of an individual as distinct from it. In this drama no scope remains for the traditional Stoic position in which people, however grim their circumstances, were still somehow able to control their own thoughts and behaviour. 'Outward things are not in my power: to will is in my power', as Epictetus put it.[1] This was still the position which W.E. Henley expressed a century back:

> Out of the night that covers me,
>   Black as the pit from pole to pole,
> I thank whatever gods there be
>   For my unconquerable soul . . .
>
> It matters not how strait the gate,
>   How charged with punishments the scroll,
> I am the master of my fate
>   I am the captain of my soul.
>
> (From 'Invictus: In Mem. R.H.T.B.')

Similarly, Taoist doctrine assumes that we can alter our own attitude to the cosmic way and thus control the actions which express that attitude, even though the way itself determines all our outside circumstances. And many people who are genuinely trapped by their circumstances have found this a workable position. By contrast, modern pseudo-scientific fatalism extends this predicament to everybody. 'We' ourselves then vanish into the machine, a process helped by a lively tradition of science fiction which has long blurred the borderline between machinery and people. Occasionally, the impossibility of seriously maintaining such fatalism leads Richard Dawkins to declare – as if he had not abolished the free, autonomous self – that 'our conscious foresight – our capacity to simulate the future in imagination – could save us from the worst selfish excesses of the blind replicators. We have the power to defy the selfish genes of our birth'.[2] This power would indeed surely be needed if the book was to deal, as is offered on its first page, with 'the deep problems: Is there a meaning to life? What are we for?'. If action was impossible, these problems could hardly arise. But hasty spasms like these cannot convince readers who have taken seriously the message of the book's opening pages.

Critics of sociobiology have not paid much attention to this bizarre fatalism. They object, and with reason, to sociobiological rhetoric. But in substance many of them, like other scientifically-minded people today, think that they are committed to physicalistic fatalism. Though for practical purposes they live their lives on the assumption that they are somehow in charge of their actions, they are unsure how to defend it. Thus the theoretical and the practical parts of their lives are radically disconnected. They regard the problem of free will as insoluble. And so it is, in the form in which it is now commonly posed. We need to rephrase it altogether.

A STRANGE STORY

I think the best way to start doing this may be to turn a much sharper eye on the supposedly 'hard' scientific view to which these people now think they are committed.

Let us, then, have a look at the steam-whistle theory of mind (usually called epiphenomenalism) which says that what happens in our consciousness does not affect the behaviour of our bodies. Subjectivity is then a causal dead end. Our experience is just an *epiphenomenon*, which means idle froth on the surface, a mere side-effect of physical causes. Consciousness is thus an example – surely a unique one? – of one-way causation, an effect which does not itself cause anything further to happen.

This doctrine has, I think, been somewhat protected by its impressive name. That is why I am using a different one, a name drawn from T.H. Huxley, who invented the idea a century back. Huxley wrote that consciousness, both in humans and animals, 'would appear to be related to the mechanism of their body simply as a collateral product of its working, and to be as completely without any power of modifying that working as the steam-whistle which accompanies the working of a locomotive engine is without influence on its machinery' (*Methods and Results*, New York, Appleton and Co., 1901, pp. 240 and 243–4). This view has since been backed by many influential theorists, notably B.F. Skinner: 'The punishment of sexual behaviour changes sexual *behaviour* and any feelings which may arise are at best byproducts. Our age is not suffering from anxiety but from the accidents, crimes, wars and other dangerous and painful things to which people are so often exposed' (*Beyond Freedom and Dignity*, Harmondsworth, Penguin, 1973, p. 20).

The fact that Skinner is writing here about feelings rather than thoughts does not make his position any more defensible. Modern epiphenomenalists do indeed often rely on this distinction between the 'cognitive' realm and a separate phenomenon called feeling, consciousness, experience or the like. But the distinction cannot work. All thought involves feeling (for instance, feelings of satisfaction or dissatisfaction with an argument). And nearly all feeling involves thought (for instance, anxiety nearly always includes the thought of some specific disturbing circumstances and possibilities). Colin Blakemore, however, has recently supported a position close to Skinner's, writing that:

> The human brain is a machine which *alone* accounts for all our actions, our most private thoughts, our beliefs. It creates the state of consciousness and the sense of self. It makes the mind . . . To choose a spouse, a job, a religious creed – or even to choose to rob a bank – is the peak of a causal chain that runs

back to the origin of life and down to the nature of atoms and molecules . . . we feel ourselves, usually, to be in control of our actions, but that feeling is itself a product of our brain, whose machinery has been designed, on the basis of its functional utility, by means of natural selection.

(*The Mind Machine*, London, The BBC, 1988, pp. 269–71; emphasis mine)

Blakemore does not explain how natural selection could 'design, on the basis of its functional utility' a capacity which has no effect on our behaviour, nor how it could then delude us into supposing that it did have such an effect. Nor does Francis Crick explain this evolutionary puzzle, though he has lately decided that he, too, has invented epiphenomenalism, calling it – in the title of his book – *The Astonishing Hypothesis.*[3]

Now I suggest that it is not helpful to repeat a doctrine that one cannot actually believe in.

Let us ask, then, what it would be like to be an actual believer in Huxley's and Skinner's kind of epiphenomenalism. (More recent kinds are not likely to be any easier to justify, because they too try to split human beings at a point where splitting does not make sense.) Here is the scenario. Somebody (our old friend A) rings up a colleague, B, to ask for help. B, however, is a real epiphenomenalist – not the ordinary hypocritical kind. So he answers at once, 'No, I'm sorry, I can't do anything for you. You see, anything that I might try to do would involve my making a conscious effort. And it is simply not possible for my conscious efforts to have any real effect in the world. I can day-dream about helping you. Indeed, I am doing so now. But that won't produce action. Though we think we are consciously controlling our thinking, that is only an illusion. Consciousness is really a separate entity from thought, an inert side-effect of real causal factors. Since you and I are both equally helpless we must submit to the vast cosmic machine. What will be, will be.'

Just a verbal matter?

How much does the oddity of this response matter? Is this – as physicalists such as the Churchlands usually suggest – just a linguistic matter, just one more surface clash between the language of science and the crude folk-psychology of everyday life?[4] If so, then we must, as Bishop Berkeley put it, 'think with the learned and speak with the vulgar'. We then assume that the vulgar have nothing of value to contribute.

If that were right, our everyday belief in the effectiveness of our conscious efforts would be something like an optical illusion or at best a rough approximation, like our belief that the sun is going up, or that the table is solid. B's fatalistic reply would then be essentially correct and scientific, though perhaps still incomplete. Our current crude concepts would then either have to be translated into these fatalistic terms or – if that proves too awkward – be eliminated as superstitious.

This, however, is not going to work. The conflict here is of quite a different order, cutting too deep for this kind of reductive reconciliation. When we change our ideas of the sun and the table to incorporate more complex background structures, we do not have to disrupt our view of ourselves entirely. We can accommodate both views inside a larger picture. But if we really believed that our conscious efforts in thinking did not affect our actions then we would no longer have any confidence in those conscious efforts at all. And people who, for some reason, actually do lose that kind of confidence stop making those efforts, including the effort to go on reasoning. They are the people who are now described as depressives.

Epiphenomenalism is at present a main obstacle that stops us thinking clearly about free will. It has not received much attention because debates on free will have tended to centre on what happens *before* choice – on the earlier causal chains that are held in some way to 'determine' what we choose. But what happens after we choose is just as important. If our choice could not affect our actions anyway, then the question whether choice itself was free would scarcely matter. Determinism, as currently explained, is not usually supposed to involve fatalism. But if determinism includes epiphenomenalism – as it is now quite commonly thought to do – then it surely does require fatalism. The only reason why this has not been noticed is that theorists proclaiming determinism usually think of it as applying to other people's lives, not to their own.

The same fashionable mistake – the same drastic oversimplification of a most complex phenomenon – is at work in both these areas, both before and after choice. *Consciousness is still not being thought of adverbially, as a mere matter of our acting consciously. It is still being conceived as a substance – a separate, supernatural entity to be sliced off with Occam's Razor. This is what stops it being accepted as a normal aspect of mental activity, an emergent capacity acquired naturally by social creatures during the regular course of evolution.* Accordingly, attempts are made to bracket the suspect entity out of the causal sequence. But this misdirected parsimony produces starvation and collapse. All our thought is predicated on the belief that when we try to decide something, and do decide it, we can put our conscious decisions into practice. A theory that ignores the difference between a normally active person's perceived thoughts about future action and those of a paralytic is not a realistic theory.

EFFECTS ON THE PRACTICE OF SCIENCE

This is true above all of the practice of science itself. Let us consider another character, C, who is trying to write a scientific article. C does not just sit still. After a time, his or her hand moves, writing or typing, and things do not end there. Still more aggressively, C fetches a book to look up a reference. Finding there something that requires checking, C then telephones the lab to arrange to do an experiment, and later in the day actually does it.

Now, we cannot treat C's conscious thought – the effort of thinking that he or she experiences – as Skinner did, as a mere result of this behaviour. The time-order is wrong. The thought is the source of the behaviour, not its outcome. Nor can the thought be seen (as it more commonly would be today) simply as a result of activity in the brain. Of course that activity is needed if thought is to occur. But the earlier thoughts are needed too. And if *we want an explanation of C's actions, the only place where we can find it is in those thoughts.* If (for instance) we ask why this researcher looked at this special page of this particular book rather than another, or proposed one particular experiment and did it in a particular way, the answer has to be in terms of its relevance to that ongoing train of thought. There is no other kind of way of accounting for these details. And relevance simply is not a property of neurones. It's a logical relation between ideas. Only if C's reasoning turns out to be crazy shall we start investigating his brain-states. And in that case they will provide explanations of the craziness, not of the reasoning.

This is why it is no use trying to study the function of consciousness by asking what would be missing if the same actions were performed by unconscious robots ('zombies') – a topic to which the *Journal of Consciousness Studies* has devoted an entire issue.[5] Many – perhaps most – of our actions are ones that would never take the shape that they do except as the outcome of conscious thought and effort – whether at the time or earlier. They are actions that had no other kind of reason to occur. Anyone who doubts this need only look at what happens – for instance in driving – on those all-too-numerous occasions when we do act absent-mindedly, without due care and attention.

Of course it is true that our actions are also shaped by a great crowd of other factors of which we are never conscious. But among these factors conscious, attentive effort has a crucial place. Removing it from the gamut of causes would be like removing temperature from one's account of climate. I have used the example of scholarly activity because it seems to me a peculiarly unanswerable case of this obvious truth, a case that perhaps is even strong enough to pierce through the screen erected by current academic nervousness about this unfamiliar ground. This case cannot be obscured – as simpler cases such as seeking food or running away from danger might – by claiming that the actions could just as well be done unconsciously.[6] An academic whose work is criticised is not likely to mount the defence that he didn't happen to be conscious when he wrote it. But the same thing is plainly true of the great mass of deliberate human activity and indeed of much animal activity too. When we try to understand it we rightly ask: 'What state of mind lies behind this? What are they trying to do? What in their experience makes them choose this course rather than another one? What are things like from their angle?'

This is not short for asking 'what is happening in their neurones?' because quite different kinds of consideration are needed for that explanation. Social and psychological explanation calls primarily, not for neurological knowledge, but for background knowledge of the social context and sympathetic understanding of aims. Nor is it short for asking 'what background of information would a logician or cognitive scientist think formally appropriate for this behaviour?' – a question which is well known to produce quite different answers. The formal information-systems that interest cognitive scientists are abstractions – collections of dried flowers drawn from the much richer and wilder woodlands that we call states of mind. Of course these information-systems are often of great interest. But they are not facts or events in the world, any more than Platonic forms are. 'Cognition' in the sense of arranging information – which zombies are supposed to be quite good at – is not an independent event or fact in the world at all. Whichever sense of that ambiguous word *information* may be involved, information-systems are highly simplified patterns, highly selective maps of certain aspects of experience.

With the railway engine, things are of course quite different. If we ask why this particular train started out at 6.15 for Bristol rather than at 9 o'clock for Torquay, it is no use whatever to investigate the details of the steam-whistle. The steam-whistle analogy was meant to suggest that conscious experience is merely a clamorous and noticeable, but irrelevant, accompaniment of action. Sometimes indeed it can be so. People can get 'carried away'. But in cases like scientific work – which is much more typical of ordinary action – conscious thought corresponds rather to the railway timetable and the intentions of the people who composed it. It is the only way in which we can *explain* those actions in the ordinary sense in which we constantly need to do so – that is, explain why they were done rather than something else. This kind of explanation does not compete with the kind which describes how the brain, and the railway engine, have to be in working order to make this kind of action possible. Both accounts are equally necessary, but for different purposes.

Might the researcher's conscious thought be only a cover-up, a post-facto rationalisation of acts already organised for it by the limbic system? E.O. Wilson made this suggestion in the opening pages of *Sociobiology, The New Synthesis*[7] but it cannot really be made to work. The limbic system is indeed important but it doesn't read books. It is simply not in a situation to recommend that a particular page should be looked at, nor to suggest what experiment will be relevant to establishing a certain argument. *What reads books is the whole person.* And since that person is likely to use different imagery, different associations, different wording on different occasions of thinking the same thought, no useful neuronal regularities are likely to attend the process. The limbic system's business is with motivation, not with working out the details of conduct.

In any case, rationalisation can never give more than a limited explanation of any conduct. Rationalisation is parasitical because it works by imitation. It must have an authentic model to imitate.

The model in this case is the normal, natural connection between conscious thought and the actions which flow from it. Without that normal connection to imitate, rationalisation could never get started.

The reason why we don't normally seek neuronal explanations for people's actions is not, then – as physicalists suggest – that neurology is still too incomplete. It is that those explanations lie in the wrong direction. They are simply not answers to the kind of questions that we usually need to ask in order to explain actions, but to questions of a different kind. The fact that there are questions of different kinds may be upsetting but it is unavoidable. We will come back to this troublesome feature of the world later.

11

DOING SCIENCE ON PURPOSE

FREEDOM AND RESPONSIBILITY

First, however, let us look again at our original question about the meaning of free will. What is really happening when a scientist, or any other reflective agent, thinks calmly about a problem, makes a decision and acts on it? If we actually believed that conscious efforts played no part in producing this action, we would have to stop taking both our own efforts and those which we now attribute to others seriously. We would then regard all actions – including the writing of learned articles – as being meaningless in the same sense as the shapes of letters that might appear by mere chance on the surface of decaying walls. ('Mere chance' here does not of course mean the absence of causes but the absence of any deliberate intention.)[1]

The actions of seriously brain-damaged patients are indeed sometimes considered as meaningless in this way. But this is because the agent's thought no longer provides the normal kind of explanation for them, so they can only be explained causally.

But, as David Hodgson has sharply pointed out,[2] it is fearfully hard to see how we could possibly extend this fall-back model across the whole range of deliberate human and animal activity, including the practice of science.

Colin Blakemore, attempting this feat, has suggested that no distinction should be drawn between deliberate and non-responsible actions:

> All our actions are products of the activity of our brains. It seems to me to make no sense (in scientific terms) to try to distinguish sharply between acts that result from conscious intention and those that are pure reflexes or that are caused by disease or damage to the brain.
>
> (*The Mind Machine*, p. 271)

Of course it is true that this borderline is not *sharp*. Common sense always recognises a wide and confusing area between the extremes. We vary carefully the degree to which we hold people responsible, allowing for endless subtle excuses and mitigations in order to prevent unfair punishment, a danger which rightly concerns Blakemore (as it did Skinner). But punishment affects only a tiny corner of the vast range involved, which is the whole scope of meaningful, deliberate action, including, as we have seen, the practice of science itself. There it is certainly considered important that people should receive credit for their writings and discoveries. These are not regarded as simple by-products of a neural process over which their authors exercised no conscious control but as testimony to those authors' deliberate thoughts and intentions. And over the whole range of human activity that pattern runs so deep that it is hard to see how it could possibly be eliminated.

By saying that he writes 'in scientific terms', Blakemore might, of course, have meant only that this distinction is not relevant for neurology. But in fact he explicitly relates his view to judicial responsibility. If it is to apply there, this argument must mean that the one-way causation – from brain to mind – excludes the reverse process. But, as we have seen, in scientific work and in countless other contexts, effective causation runs the other way too. Like most causal pathways, this one carries two-way traffic. That is what makes the distinction between responsible and irresponsible action a real and necessary one.

TROUBLE OVER FOREKNOWLEDGE?

People defend these wild positions because they assume that explanation must always follow a single path. They rightly do not want to interrupt the neurological story by inserting occasional bits of conscious thought into it. And they do not see how a parallel mental story can be anything other than a dangerous rival. But the neurological story is always essentially incomplete because the brain is not the whole person. Nor (of course) is it a sinister puppet-master who has devised the rest of the person as its helpless plaything. Our brains are parts of us, useful though rather complicated bits of meat packed inside our skulls. No doubt the biological account of brain-processes is continuous in its own terms. But it takes for granted the context of a wider process which has to be observed from other angles as well.

A misleading idea that constantly intrudes here is the picture of the neurones as forcing the thought. They no more force it than a plant's genes force the plant to grow. The brain supplies part of the means of thought just as the genes supply part of the means for the plant's growth – a part which, incidentally, has been considerably exaggerated during recent decades.[3] It is the agent as a whole that acts. And that agent acts freely in so far as he or she is free from various forms of what Spinoza called human bondage – for instance neurosis, paranoia, self-deception, ignorance, cowardice and other genuine defects. These are the kind of factors that threaten freedom, not the normal influence of our bodies.

Is this freedom compromised by the idea that our thoughts are predictable? Here again, the trouble surely stems much more from the fatalistic superstition which attributes control to a powerful fate-figure than from the possibility of prediction itself. If one's ideas have been secretly inserted into one by a Svengali, then one is certainly in trouble. More generally, too, the idea that any human expert knows more about my future actions than I do is frightening because that person then has undue power over me. But the idea that what makes us predictable is that our own bodies are alien Svengalis exercising power over us is not very sensible. For instance, take someone like the ex-slave-woman Epicharis, who (as Tacitus tells us) had been tortured for a whole day by Nero's police to make her betray her fellow-conspirators. If neurologists had been given a full account of her brain-states, would they have been able to predict whether she would eventually give way or would somehow manage to kill herself in order to prevent this betrayal, which is what (to the astonishment of all) she actually succeeded in doing?[4] Nobody knows whether such a prediction could be made, even (as they say) 'in principle'. But, if it had been made, it would not have had any more authority than the prediction which her friends might have made on totally different grounds. Neither story is anything like complete. And neither kind of outcome need have involved interference by a power or agent other than Epicharis herself.

This social aspect – this idea of a power-relation producing predictability – shows our central selves as separate beings forced, against their will, to act as the slaves of our bodies. That idea is commonly associated with determinism. But determinism is not a clear doctrine. In fact the word *determine* is awkwardly ambiguous. It is sometimes used in the impersonal sense in which three points determine a line. And here it has little force if applied to causality because (in real life as opposed to mathematics) we never do deal with just three points but always with an open, indefinite range of causal factors. The stronger sense of the word, however, is that in which a general can, by choosing, determine the fate of a private. The word then means 'control'. We often do import that dramatic sense into our talk of causality. But to do so is (I am suggesting) superstitious fatalism.

DUALISM AND METAPHYSICAL WARFARE

To sum up, then, so far – epiphenomenalism, even in its modern forms, is not compatible with the central structure of concepts which make possible the whole business of attentive human action, including science. The clash cannot be smoothed over on the reductive model of our language about the rising sun. B's excuse is not an incomplete bit of science, containing some useful half-truth. It rests on no facts, scientific or otherwise, merely on epiphenomenalist doctrine. And that doctrine is not itself a fact but a piece of metaphysics, part of the Cartesian conceptual map on which we still too often organise our ideas.

Descartes designed his dualism as a protective railing to save the tender shoots of physics from being eaten by alien organisms such as theology. But today the mighty tree of physics does not need these protective railings. In fact, physics itself has recently broken through them by bringing into quantum mechanics questions about observers. Despite attempts to confine them to the recording apparatus, these questions are generally understood as questions about subjects.

Meanwhile the social sciences, which have grown up along that awkward Cartesian frontier, have long been in trouble trying to deal with a customs barrier which splits them down the middle. This unrealistic division has also gravely distorted medicine, especially psychiatry, and our understanding of evolution. In all these contexts, we need to retrain ourselves to grasp that we are not forced to choose between two different kinds of units, mind and body. The unit is the whole person.

Descartes left his two substances unrelated except for the

external relation of both to God. Inevitably, each province then tried to unify the whole by imposing its own methods on the other sweepingly, by abstract logic, by inventing vast, rigid rationalistic schemes. At first, these takeover bids came largely from the mental side. Idealists from Leibniz to Hegel launched impressive systems. But these systems never managed to digest the physical sciences. And after Hegel unfortunately proved *a priori* that there could only be seven planets just before the discovery of the eighth, their popularity waned.

At that point, physicalist counter-imperialism seemed to have prevailed. Descartes' project of reducing the other sciences to mathematics via physics was so marvellously effective on its own ground that by the nineteenth century many people hoped to extend it over all the rest of thought. It was expected that the close relation of physics to chemistry could eventually be duplicated all over the intellectual scene. All reputable forms of thought would then be piled up in a linear series with − so to speak − no logical space left around them or between them. Explanation would be a continuous, one-way process moving always towards the safe physical terminus.

That was the programme which Auguste Comte confidently proclaimed as part of the gospel of positivism, and which Victorian intellectuals welcomed as a way of cutting out the metaphysical extravagances of religion. The only question then was how to spell out these simplifying, reductive explanations in detail across the full range of the humanities − across, for instance, geography, history, logic, law, musicology, linguistics and ethics as well as the nascent social sciences.

This proved a lot harder than Comte expected. At first, theorists tried to do it by building ambitious systems which they called scientific but which had no real connection with the methods of the sciences. Herbert Spencer gave his life to this enterprise, but his efforts have not worn well. Others produced quasi-scientific laws of history such as Marx's and Spengler's, or bold psychological systems such as Freud's. Behaviourism itself is an ideological system of this kind, as indeed Comte's positivism itself was. They are all really best seen as moral and political visions rather than as extensions of any science.

In time, the sheer number and variety of these schemes, as well as their actual faults, showed that they could not command the kind of consensus needed for a science. Early in the twentieth century, then, serious supporters of the reductive project turned from imperialistic conquest to isolationism. Rather than try to turn history, poetry, ethics and the rest into parts of science, they worked to restore scientific purity by shutting these strangers out. Finally, following Popper, they narrowed the definition of science so far that it excluded, not just Marx and Spengler and Freud, but much of the social sciences as well. (By an unlucky oversight they also shut out Darwinian evolutionary theory, which had to be brought back as an awkward exception.[5])

It was during this purist epoch that behaviourist ideology raised its wall to protect 'science' from any reference to conscious subjects. Epiphenomenalism has been the barbed wire on top of that wall, the barrier which has fenced the scientific imperialists – the people who still want science to conquer all other studies – into their present awkward corner. On the one hand, they hold that science is the only reputable form of thought, everything else being either religion or 'pseudo-science'. On the other hand, they now define science so narrowly that this story cannot possibly be true.

THE RETURN OF THE SUBJECT

The behaviourist project of confining the whole scope of science to *objects* has, then, visibly become hopeless. All over the sciences themselves, as well as elsewhere, serious questions are arising which involve people (and indeed other animals too) as subjects as well as objects. These questions cannot be dealt with by methods which have been carefully designed to suit objects alone.

'Objectivity' understood in that sense has turned out to be a will o' the wisp. Thus there is not just one problem of consciousness nor could there be a single 'science of consciousness' to deal with it. What we have is a whole range of previously suppressed problems clamouring for our notice. We do not need just to find the neural basis of sentience but to unpack the whole rich and puzzling notion of consciousness as it normally works in our thought – including, of course, its value-aspect, the reasons why we think consciousness important.

Today's scientists, however, have been sternly trained to think that all rational thought has to follow only the pattern designed for studying objects. Subjects, if they are to be rationally considered, must therefore be forced onto that same Procrustean bed. For example – if we ask how it is that C's thoughts can produce acts appropriate to them (such as checking a given reference and setting up a given experiment) they feel that they ought to look for universal causal laws governing such processes, laws modelled perhaps on the laws of physics, as David Chalmers proposes. ('Where there is a fundamental property, there are fundamental laws.'[6]) But that kind of law cannot be found here, because appropriateness is a concept that works quite differently from the concepts of physics, though naturally it is one that physicists must continually use in their own thinking.

The laws of a particular science only operate within it, not between it and its conceptual neighbours, where wider structural considerations have to be used. Purists therefore conclude that the connection between our conscious efforts and our actions is unreal and must simply be denied. That connection is, however, the best-attested causal link in our whole experience. Indeed it may well be the model on which we originally conceive all other causal links. Yet for a time denial of it was proclaimed to be a necessity for science. People who were unhappy about leaving human behaviour completely unexplained in this way were told to have faith that the strict causal laws which are needed for it would one day be provided – by neurology.

But why neurology? Innumerable kinds of causes and conditions are necessary to bring about C's action – social causes, educational causes, economic, political, climatic, nutritional, genetic, medical causes, besides the subjective and neural angles that we have noticed. Each of these sets of causes may partially explain the action. But none of them invalidates the others; there is room for them all. The question which of them counts as the explanation in a given case depends entirely on our interests, on what we want to find out.

Current notions of determinism, however, allow specialists in any of these topics to feel that the whole logical space is theirs – that they alone possess the explanation. Thus we get the odd spectacle of many competing determinisms – genetic determinism, economic determinism, neurological determinism and so forth – where the claim seems to be that a single discipline has finally found the engine which runs all the other causes. That is another trouble about current notions of determinism which needs to be sorted out. On large topics, this kind of tribal narrowness is always disastrous.

ESCAPE ROUTES

A surprising number of current theorists have got stuck in this corner. One obvious way out is to use a less narrow and arbitrary notion of causality – a notion which can intelligibly describe a wider range of connexions, especially a wider range than is found in physics or chemistry. Two very helpful enquirers, Rom Harré[7] and John Searle,[8] have already worked out this escape route in constructive ways. They point out that the current use of the word 'cause' can accommodate this extension quite easily, as we see when we ask about such things as the causes of unemployment, or of current discontent, or of the French Revolution, where there is no question of invoking universal laws like those of physics. There is, in fact, always a deep connexion between the notions of cause and reason – a connexion which has to be grasped if we are to make sense of the relation between subjects and objects. The word cause has always been regularly used for reason, motive, source, explanation or even agent quite as often as for 'occasioning event'. Indeed, the *Oxford English Dictionary* cites from Thomas Reid, writing in 1790: 'In the strict philosophical sense I take a cause to be that which has the relation to the effect which I have to my voluntary and deliberate actions.' Similarly, a notice on my local buses quite properly tells me not to speak to the driver without good cause. The recent narrowing of this word to exclude this sense is not compulsory and is proving unhelpful.

It should be noticed, too, that not all explanation is causal explanation in any sense. For instance, mathematical explanation involves no reference to time, nor, more generally, does any relation between reasons and conclusions. There are many other kinds of explanation besides causal ones because there are many other kinds of question that can be asked.

COMPLEXITY IS NOT SURPRISING

This diversity of questions is a central topic of this book. There are many other ways of connecting different patterns intelligibly besides trying to force them into a uniform mould. Recognising their diversity is not a licence for confusion but a sane and effective policy.

Ever since the seventeenth century, however, our culture has made it hard for us to see this. The reductive pattern has been so successfully used to connect the various physical sciences that it always seems seductive. But in many areas it has now reached the limits of its usefulness and in others it has always been inappropriate. It is the pattern that tells us, when we encounter several different ways of thinking, to arrange them in a hierarchy, a linear sequence, running from the superficial to the fundamental, which occupies the whole logical space available for explanation. The more fundamental thought-patterns are then called *hard* while the upper layers are *soft*.

This tactile metaphor is rarely explained, but it is clearly meant as a compliment. (People who exalt hardness in this way are evidently not thinking about the possible hardening of their arteries, nor about the inconvenience of drinking ice instead of water.) The upper or softer layers are then ranked as relatively superficial because they do not give an ultimate explanation. They are considered amateurish, non-serious – as Berkeley put it, the property of the vulgar, now called folk-psychology. They are makeshifts to be used when the real scientific account isn't available, or when it is too cumbersome to use. In fact, they are just stages on the way down to the only fully 'mature' science, which is physics.

The metaphor of levels, which is often used to describe the relation between these various ways of thinking, can seem to endorse this one-dimensional pattern. But it has never become clear how the various kinds of disciplined reasoning involved in non-scientific branches of thought such as history, logic, law, linguistics and mathematics could be piled up as stages on the way to any physical science. And, as we have seen, it is still more obscure how the practical thinking that we use in active life could ever be so treated.

That is why I proposed, instead of this linear pattern, the many-maps model of the relation between different kinds of question, designed to show how these different enquiries cut into the cosmic cake, so to speak, from different angles, revealing different patterns. Cake-cutting too is a better image than talk of levels, because it does not suggest a hierarchical linear series. Or again, we could think of looking in through different windows at what happens in that aquarium. Or of asking various kinds of questions about the Chinese vase. All of these are rational procedures. And all of them give us some chance of recognising the actual richness of life.

People sometimes say that the human brain is the most complex item in the universe. But the whole person of whom that brain is part is necessarily a much more complex item than the brain alone. And whole people can't be understood without knowing a good deal both about their inner lives and about the other people around them. Indeed, they can't be understood without a fair grasp of the whole society that they belong to, which is presumably more complex still. Like the Chinese vase, in fact, human beings are not simple items.

It is surely not surprising, then, if our understanding of something as complicated as this is necessarily partial, divided and incomplete. We ourselves are not pure, omniscient minds standing outside the universe and equipped to find its ultimate structure. We are evolved animals, working under difficulties from inside the system. We use a haphazard mix of faculties which are not fully unified, but which give us various different sorts of useful light, so we had better try to use them realistically.

12

ONE WORLD, BUT A BIG ONE

WIDER HORIZONS

'Explanations', then, come in various kinds. They vary with the needs that call for them. 'Explaining things' does not necessarily mean fitting them into existing scientific schemes. It means placing them on the wider map of our thought. The current need to 'explain consciousness' does not arise mainly out of curiosity about its physical or evolutionary causes. It comes from a wider uncertainty about its place in the general scheme of things, both within our own lives and in the universe as a whole.

During much of the twentieth century, naive dogmatic materialism suggested that this uncertainty did not matter. No questions arose here because consciousness was a trivial matter with no significant place in the world at all. Behaviourist psychology tried resolutely to ignore our subjective experience altogether. The failure of that experiment has been a crucial epoch in our thinking, whose consequences we are now trying to deal with. Scientists have responded to that failure by trying to give consciousness reputable standing. At present they chiefly hope to make it 'scientific' by squeezing it in at the margins either of neurobiology or of some branch of theoretical physics.

But we have seen that an item which plays as central a role in our life and thought as consciousness does is too large to be 'explained' by being annexed to a particular physical science in this way. Nor does it need to be. The difficulty that we feel about consciousness is not a local one that can be resolved by relating it directly to physics. It is one of relating our whole inner and outer viewpoints – of finding a context in which the subjective and objective aspects of life can be intelligibly connected. To some extent this is a real, unavoidable, general difficulty which can only be handled by constant adjustment in our lives. This adjustment is attempted by all cultures and is a topic which so far is usually viewed as the business of serious literature, of history and of philosophy rather than of the physical sciences.

That topic has, however, lately been made to look much harder than it is by being conflated with the demand for physical explanation, and by the surprising notion that this kind of explanation would always be more 'fundamental' than all others. This really is a mistake. Physical explanations are only fundamental for physics. Other kinds of question need other kinds of answers. The main difficulty is still to identify the exact question we are asking.

THE IMPORTANCE OF SOCIAL FACTS

Is matter somehow more real than mind? Is physical explanation always more fundamental than other kinds of explanation? These odd questions, on which we touched in the Introduction, surely lurk in the background of current debates about consciousness. But can reality be the kind of thing that has degrees? Can things be more or less real? And again, does it make sense to talk of one enquiry as more fundamental than another until one has explained 'fundamental for what?'.

I would like to approach these large issues by looking at the admirably clear discussion which opens John Searle's book *The Construction of Social Reality*.[1] Searle sets out a manifesto that lights up sharply the troubles that have lately distorted this topic. He writes, 'We live in exactly one world, not two or three or seventeen'. And that is surely right. But then comes the difficulty. Searle goes on:

> As far as we know, the fundamental features of that world are as described by physics, chemistry and the other natural sciences. But the existence of phenomena that are not in any obvious way physical or chemical gives rise to puzzlement ... How does a mental reality, a world of consciousness, intentionality and other mental phenomena, fit into *a world consisting entirely of physical particles in fields of force?*
>
> (p. 1; emphasis mine)

He rightly goes on to point out that this question is not an isolated one dealing only with hidden, private experiences – aches and images and qualia. It opens issues of a quite different order of magnitude, namely, the whole extent of social phenomena. We must go on to ask:

> How can there be an objective world of money, property, marriage, governments, elections, football games, cocktail parties and lawcourts in a world that consists entirely of physical particles in fields of force, and in which some of these particles are organised into systems that are conscious biological beasts, such as ourselves?
>
> (p. 1)

As he shows, in order to deal with these topics we have to be *objective* – that is, fair, honest and methodical – about the whole range of the subjective. We have to treat seriously, not just a few detached aspects of our consciousness, but the whole range where subjective experience affects objective facts – facts which are a real part of our single world. Since much of our subjective life is unconscious, this greatly enlarges the ground to be covered, though the conscious part, the lit centre, is still a crucial aspect of it.

Social institutions such as money, government and football, which have played little part in debates on this matter so far, are forms of practice shaped and engaged in by conscious, active subjects through acts performed in pursuit of their aims and intentions. They can therefore only be understood in terms framed to express those subjects' points of view. Physical analysis of the material objects involved may be quite irrelevant. Money, for instance, notoriously cannot be defined as a particular material thing or stuff. It gets made out of many kinds of stuff or even – as seems to happen at present – out of mere credit. Money is whatever people agree to use as a medium for exchanging goods. This has radical consequences for the project of explaining it. Someone who asks *what money is*, or who wants to understand its workings better, needs to know about the wishes and intentions of people wanting to exchange goods. This is the only way of explaining money. Physical analysis of coins or notes, or of the neurones of the people using money, would be useless and irrelevant for this explanation.

EXPLANATIONS VARY

Explanation, then, is not a standard item. It is whatever information or reasoning will solve the particular problem that is causing trouble at the time. This is a central point for the project of 'explaining consciousness' because many people discussing this assume that asking for an explanation of it is simply asking for its cause – for a physical condition that produces it. They therefore look for these antecedents – usually either in evolution or in neurology but sometimes in quantum physics – and they are mystified to see why anybody should look anywhere else.

The approach which looks always for the cause is of course very often the right one. It is suitable when we are trying to 'explain' some phenomenon such as global warming, where we are already confident that we grasp an effect adequately and the cause is then the next thing that we want to find. But the project of explaining money, or elections, or time, or marriage, or football, or grammar, or art, or laughter, or gambling, or the Mafia, or post-structuralism or the differential calculus is not like this at all. Here, what we need is to know more about the workings of the thing itself. Similarly, it is often said that DNA 'explains life' because it fills an important gap in the casual story about how life is reproduced. This gap-filling is indeed welcome. It is needed because it eliminates vitalist speculations about a special quasi-physical force or stuff-of-life which might have done this particular job. But the discovery of DNA tells us nothing about the peculiarity of life itself while it is actually being lived. And this peculiarity is often what puzzles us. We will come back to this issue in Chapter 21.

It is interesting that biologists are at present just as frightened of attempts to analyse the concept of life as psychologists were, until lately, by any similar attention to the concept of consciousness. Standard dictionaries of biology simply do not have an entry for this word. Writers who occasionally mention it usually either say that it is indefinable or refer briefly to a few formal properties such as homoeostasis, replication and complexity. But what sort of complexity? And how would we get on if we confined our understanding of money to a similar set of formal properties?

CATEGORY PROBLEMS

The reason why words like life and consciousness are awkward is that they lie at the peculiar level of generality where we have to change gear at the boundary of a category. They mark the frontier of a whole logical type. The same kind of trouble arises over terms like time, necessity, chance matter, life and reality. For these large concepts, dictionaries notoriously start to revolve in a circle of near-synonyms, dodging the philosophical problems. Even game is a much harder word to define than football, since the idea of play is itself a distinctive and somewhat puzzling category of thought.[2]

'Consciousness' is not one among a class of parallel instances as football is one among games. It is a term used to indicate the centre of the subjective aspect of life. Understanding such a word means relating that aspect fully to the other aspects. And this business of relating is – like medicine – inevitably an art rather than a science, though of course sciences can sometimes form a very important part of it.

Of course, when these awkward words which signal category-boundaries have to be used in a particular science, they can be given sharp, limited definitions to suit immediate purposes. Thus, a science which only wants to account causally for the physical conditions of consciousness can perfectly properly do this by using a vague, minimal definition of consciousness, thereby investigating what David Chalmers has called 'the easy problems'.[3] But the results may have little relevance to the wider difficulty of fitting together the various aspects of our world. The problem here is the vast one that each of us has in trying to relate our own individual experience sensibly to that of others and to the official views of our society. This problem is a central concern of serious literature and also of philosophy. It has often been held not to concern physical science. The pioneers of modern science deliberately withdrew from it in the seventeenth century, confining their attention to physical matters, and the subsequent success of the physical sciences is generally thought to have owed much to this traffic-regulation.

However, once consciousness is admitted to be real, the brute causal regularities studied by a narrower perspective do not satisfy all scientific enquirers. The scientific spirit looks also for an understandable connection, a reason why the effect is suitable to the cause. If this can be found at all in the case of consciousness – which is still not clear – the search for it must certainly involve reference to a much wider context which takes both aspects seriously as a whole. This kind of wider reference is something that current specialisation makes extremely hard for academics. For a start, it is surely going to include the social context which Searle mentions and also, beyond that, the whole context of earthly life.

As for life, I suspect that, twenty years hence, biologists may be concerning themselves with 'the question of life' just as vigorously as the rest of us now are with problems about consciousness. As Lynn Margulis and Dorion Sagan have lately pointed out, it really does not pay to neglect one's central concepts.[4] In both these cases, one question that enquirers often have in mind is 'what kind of importance or significance these things have, what makes them matter so much?'. The weight that we attach to the difference between being alive or dead, conscious or unconscious, for others as well as ourselves, is not just a chance matter of our individual fancy but a part of the concept.

Importance decides what we select for attention, which is why science cannot be 'value-free' in the sense of neglecting it. The assumption that consciousness was unimportant was what led the behaviourist psychologists to their bizarre and unworkable conclusions. If we now decide to take the matter seriously we shall have to use different methods. That is how we come to be led into much larger philosophical problems such as free will and personal identity – problems which deeply affect our view of the kind of creatures that we ourselves are. These questions are primarily about finding the best way to think rather than about strictly factual matters. Though, therefore, they concern scientists and raise many scientific questions, they do not seem to be in the narrow contemporary sense purely scientific questions. I am not sure whether the current idea of 'a science of consciousness' is meant to define a territory parallel to that of, for instance, 'a science of digestion', but if it is, it is not going to work.

MAPPING THE BACKGROUND

What sort of explanation, then, is needed here? In the case of the social items which I listed such as money and football, explanation involves mapping the structure of each and showing how that structure connects it with the surrounding conceptual areas of life as a whole. Among those areas the participants' aims and intentions form a crucial core. We always need to ask 'what are these people trying to do and why?'. This gives us, not a causal story in the sense now expected in physical science, but a historical story. B.F. Skinner dismissed 'those collections of personal experiences called history' as amateurish and inadequate because they did not use the methods of a physical science.[5] History in fact works by the disciplined use of multiple analogies and the careful scrutiny of evidence. These methods are not antiscientific. They are as rigorous, as necessary and as reputable as the methods of the sciences. Indeed they are needed and used in important areas of science itself, such as cosmology and the study of evolution, because those areas deal with single, unrepeatable historical processes, exactly as the study of past human life does.

The kinds of explanation we need vary, then, according to the nature and state of our topic. Different kinds of concept need to be explained in different ways, ways which track the current gaps in our ability to handle this particular subject-matter. Like explorers in a territory where few places are yet known, we try to find connections between our existing insights. And, like those explorers, we often find reason to distrust the maps which have been used so far.

The enterprise of relating different aspects of life is, however, trickier than that of relating lakes and mountains because there are more ways in which they might be related. It is sometimes very hard to be clear about just what it is that we have got right and what it is that we still do not understand. That is the kind of trouble we are presently having about consciousness. We really need new concepts that will bring the whole person into focus again. We need to correct the unrealistic division which has long distorted our thought, and which is now being constantly widened by the image of the mind as a computer-program, carrying the sharp, simple dualism of software and hardware.

THE ROOTS OF DUALISM

This claim about variation, this declaration that kinds of explanation differ and new ones must be found, is not a shocking manifesto for irrationalism. It does not mean that just anything can count as an explanation. It merely means that the one world in which we all live is complex and that our powers of thought are, luckily, suppler and more versatile in dealing with it than some philosophers have supposed. The many ways of thinking that we possess, and are always developing, are not rival alternatives. They form a potentially coherent set of tools which we always learn to use harmoniously to some extent (as carpenters do) and whose various capacities we can come to understand better with careful practice.

However, ever since the time of Descartes the ambition to simplify thought by reducing it to one great underlying form has been strong in Western philosophy. Descartes' faith that the world itself is built on a single, fundamentally simple pattern has only lately begun to be questioned.

His dilemma still deserves our attention because it is essentially the same one that has cropped up again today. It has done so now because we can no longer live with the inadequacies of his simple solution to it. Descartes was the first great thinker who was struck by the peculiarity of inner experience; the full, strange specialness of the subjective viewpoint. Too honest and clear-headed to try to sweep this specialness under the carpet by saying that experience was somehow unreal, or was not important enough to be taken seriously in our understanding of the world, he saw that this personal experience was in a way the centre of all importance. He therefore made it the starting-point of his enquiry, remarking, *Je pense, donc je suis* – I think, therefore I am.

But *what sort of a thing is this I?* (This is where the real difficulty starts.) Descartes was (again) too honest and clear-headed to think that it could be tucked in somewhere among the subject-matter of physical science – as many people apparently want to do today. Ahead of his time, Descartes fully appreciated the dawning science of people like Galileo and Harvey. He was eager to display its importance. But he understood that the success of this modern science was due precisely to its sharply limiting its subject-matter and excluding from it concepts such as purpose which presuppose a particular point of view – concepts which (as he saw) had blocked and confused earlier physical theorising.

That is why Descartes divided mind from body radically and ontologically – by declaring that they were unrelated kinds of substance, linked only by God's external mediation in a perpetual miracle. In this way reason, which thinks about both items, could still be viewed as a single faculty although it used different methods for these two different purposes, because the difference of concepts was due to the difference between the two subject-matters. It was, then, still possible to hope that, on both sides of this gap, reason would soon produce beautifully simple explanations. This notion of a deep split in reality was the background that was taken for granted in the first days of modern science.

THE FAILURE OF UNIFICATION BY CONQUEST

Many people, however, always suspected that the idea of such a radical division was unworkable for just the reasons that are now bringing forward the 'problem of consciouness'. It is not really possible to keep up a reason-proof barrier between subjects and objects, between inner and outer, between thought and things. We continually need to think about relations between these two elements in the world, so we cannot effectively hold them apart. Their relations are essential sinews of our thought. Explanation cannot work without continually crossing this supposed gulf. As Searle rightly says, we live in just one world.

How then could the gap be mended? As we have seen, human pugnacity ensured that the first method which occurred to most theorists was conquest, and again this was attempted by the extreme methods of ontology, through an ambitious, sweeping metaphysical colonisation of matter by mind or mind by matter. From the seventeenth to the nineteenth century, philosophers fought to establish that either matter or mind was the one basic substance so that the corresponding form of explanation was equally basic for all thought – a project which grew more and more awkward till it exploded and sank in its final form, the Marxist dialectic.

13

A PLAGUE ON BOTH THEIR HOUSES

THE VISION OF MATTER

What then were these warriors saying? To summarise crudely, on the idealist side it seemed that matter must be shown as simply a form of mind, a mere logical construction out of sense-data (Leibniz, Berkeley, Hume, Hegel, Ayer). On the materialist side, it seemed equally clear that mind was only a form of matter, that 'the brain secretes thought just as the liver secretes bile'[1] (Hobbes, Laplace, La Mettrie, Marx, Skinner). T.H. Huxley managed to embrace both kinds of reduction at once, but this was an unusual achievement.[2]

Unfortunately, despite some serious attempts at subtlety in both armies, this ontological warfare was too crude a practice to sort out the difficulty. Neither alternative can really be made to work. Both, indeed, have their strengths and both have had their epochs of success. Anyone who has not yet felt the force of the idealist position need only read Hume and Berkeley to discover that it is quite as easy to start dissolving away matter as it is to start abolishing mind. But after the first few moves both enterprises run into grave difficulties. Recently, materialism has certainly claimed the field. Many people today still think it a meaningful, reasonable doctrine. To the contrary, I want to say flatly that it is no better than its opponent and probably worse – since a world without subjects is even less conceivable than a world without objects. And if one gets rid of one alternative one must equally get rid of the other. The whole ontological quarrel is mistaken.

The current credulity about materialism is understandable because – quite apart from the attractions of the traditional warfare against God – the way in which the dispute has lately been thought of makes it seem unavoidable. For some time the debate has looked like one taking place within the physical sciences themselves. The combatants have tended to stay close to Descartes' idea that the two rival elements were different substances. Unless one takes the intense logical care that Aristotle used, this simply sounds like two kinds of physical stuff. Disputants – especially on the materialist side – have therefore had the impression that they were asking something close to the pre-Socratic question 'what physical stuff is the world made of?'.

Mind then naturally evaporates because it looks like a kind of gas, a gas which is certainly not recognised by modern chemistry or physics. Nor do things get better if, instead of a substance, mind is treated as a force closely comparable to physical forces. Again, there is simply no room for such an extra force on the physical table or anywhere near it. This is just the mistake which the nineteenth-century vitalists made when they tried to insert a quasi-physical force or entity called 'life' as a factor on a par with those already considered by physics and biology, instead of pointing out that the extreme complexity of living things called for quite different forms of explanation.

CONSCIOUSNESS AS AN HONORARY PHYSICAL ENTITY

The current wish to take consciousness seriously has, however, led people to hope that they can legitimate it by finding it a place on the borders of physical science, without repeating the vitalists' mistake of trying to locate such an item inside it. Thus Gregg Rosenberg suggests that 'the irreducible character of experience implies that *fundamental natural laws* are governing it, laws on the same level as those governing properties such as mass, motion and gravity'.[3] This is a variant of David Chalmers' suggestion that we should avoid reductionism by taking 'experience itself as *a fundamental feature of the world, alongside mass, charge and space-time*'. (Both emphases mine.) Chalmers remarks with satisfaction that, if his view is right,

> then in some ways a theory of consciousness will have more in common with a theory in physics than a theory in biology. Biological theories involve no principles that are fundamental in this way, so *biological theory has a certain complexity and messiness about it*: but theories in physics, insofar as they deal with fundamental principles, aspire to simplicity and elegance. The fundamental laws of nature are part of the basic furniture of the world, and *physical theories are telling us that this basic furniture is remarkably simple*. If a theory of consciousness also involves fundamental principles, then we should expect the same.
>
> (David Chalmers, 'Facing Up to the Problem of Consciousness', *Journal of Consciousness Studies*, vol. 2, no. 3, 1995; emphases mine)

Physics-envy could hardly be more touchingly expressed. Life, however, is essentially messy. The trouble about this kind of proposal is an extremely interesting one. Rosenberg wants to insert his new laws 'on the same level' as those of physics. This is evidently because he thinks physics is the ground floor, the bottom line, the slot for the ultimate and most important classificatory concepts, the only place for categories. So does Chalmers. And both rightly think that 'experience' has that kind of importance because it is a category-concept, a concept too bulky to be accommodated in a mere annexe on the edge of neuroscience. They therefore want to insert it as a physical category among the largest and gravest kinds of concept that they can think of.

But this place cannot be found by annexing it to physics, any more than to neurobiology. Physics is an immensely specialised science. Its basic concepts are most carefully abstracted, neatly shaped to fit together and to do a quite peculiar conceptual job. They cannot accommodate an honorary member of a different kind. The meaning that concepts like 'mass' have in physics bears little relation to their everyday meaning. Trying to add a rich, unreconstructed, everyday concept such as consciousness to this family is like trying to add a playing card to a game of chess – or perhaps more like trying to put down a real queen or knight on the chessboard. These new items are of a different logical type. They need a different type of context. They do not belong in this game at all. They can't be 'on the same level'. The category-difference is too great.

Anyone who thinks that physics could conveniently build on this kind of annexe should notice that, if it did, plenty of other concepts would have as good a claim to occupy it as consciousness does. What, for instance, about *substance, necessity, truth, knowledge, objectivity, meaning, communication, reality* and *appearance, reason* and *feeling, active* and *passive, right* and *wrong, good* and *evil*? What indeed about *life*, which has only been excluded because people today don't want to look at it? All these are basic categories of our thought. If physics were enlarged to accommodate all of them it would become continuous with the philosophy of science and through this with the central areas of metaphysics. This might not have worried some of its greatest proponents, from Galileo to Einstein. Perhaps indeed it needs this kind of outward connection. But such a move would run quite contrary to the Popperian limitation of the scope of science which most scientists seem to accept today.

The physical 'level', then, is not the meeting-place of all thought, not a set to which all really important concepts have to belong. There is no need to expect that other crucial areas of thought will turn out to be governed by universal 'natural laws' comparable with those used in physics or directly relatable to them. The quest for such quasi-physical laws governing history and the social sciences, a quest which was eagerly carried on by theorists such as Herbert Spencer, Toynbee, Spengler and Marx, has turned out disastrously misleading. And it has done so largely because it was guided by blind imitation of physics rather than by attention to the needs of its subject-matter.

However, the phrase 'fundamental natural laws' is, in current usage, firmly stuck to the laws of physics. This expresses a convention which equates *nature* with the abstractions studied by physics. Rosenberg wants to unstick this phrase slightly and to widen its scope somewhat by revising the idea of nature to take in consciousness. This widening project is surely right as far as it goes. But it has to operate on a far bolder and more drastic scale. Explaining the whole of nature is not a linear process directed downwards towards a single set of explanatory concepts. Thought is not a neat pile of bricks in which each is supported only by the one beneath it. Thought is not governed by this kind of gravitation: its connections go in all directions. The gravitational habit of explanation, now called 'foundationalism', is, as has lately become clear, quite inadequate.

THE VANISHING SUBJECT

Chalmers' and Rosenberg's suggestions seem, then, to be one more example of the kind of mistake which has been dogging attempts to find a place for mind somewhere among the stony fields of matter ever since metaphysical materialism became the dominant fashion in the mid-nineteenth century. People have been credulous about materialism because at a deep level they still assumed that – in however sophisticated a sense – we had to look for a single fundamental stuff, a substrate which would provide a universal form of explanation. In spite of advances in physics which ought to have undermined this pattern, the image of matter as the *hyle* or wood out of which things like tables are made persisted, and explanation in these terms inevitably made mind look like some kind of illusion. As we have seen, when the brain has finished secreting thought, its product has no weight, nor has the brain itself necessarily got any lighter. And again, there is no gap in the physical forces working on the brain which might leave room for mind as an extra force. Perhaps, then, consciousness was, as J.B. Watson sometimes put it, simply a myth?

This is the exciting terminus to which the behaviourist psychologists triumphantly drove their train at the end of the nineteenth century. Life there proved, however, so strange and puzzling that nearly all the passengers (including Watson himself) quickly moved back from it and began travelling round the various neighbouring stations on this same branch line, looking for compromise positions. They are still doing so today.

TWO STANDPOINTS, NOT TWO STUFFS

What is needed, however, is to avoid ever going down that branch-line in the first place. These two aspects of life are not two kinds of stuff or force. They are two points of view – inside and outside, subjective and objective, the patient's point of view on his toothache and that of the dentist who studies it. The two angles often need to be distinguished for thought. But both of them are essential and inseparable aspects of our normal experience, just as shape and size are inseparable aspects of objects. The dentist is aware of the patient's pain as a central fact in the situation he studies, and the patient, too, can to some extent think about it objectively. Indeed, dentists can become patients themselves. The only kind of item that has to exist in the world in order to accommodate these two standpoints is the whole person, the person who has these two aspects. Ontology has to accept that person as a single, unbroken existent thing.

Virtually all our thought integrates material taken from the two angles. As Thomas Nagel points out in an excellent discussion of these two viewpoints, we never normally take either position on its own. Instead, we constantly move to and fro between them, combining material from both:

> To acquire a more objective understanding of some aspect of life or the world, we step back from our initial view of it and form a new conception which has that view and its relation to the world as its object . . . The process can be repeated, yielding a still more objective conception . . .
>
> The distinction between more subjective and more objective views is really a matter of degree, and it covers a wide spectrum . . . The standpoint of morality is more objective than that of private life, but less objective than the standpoint of physics.[4]

Thus we combine elements derived from the two angles in various ways that suit the different matters that we are discussing, ways that differ widely according to the purpose of our thought at the time – much, perhaps, as we combine visual and tactile data in our sense-perception.

As Nagel points out, objectivity is not always a virtue, nor is it always useful for explanation. It is only one among many ideals which we have to aim at in thinking. In many situations an increase in detachment can be a cognitive as well as a moral disaster. This is, of course, most obvious in private life, which (it should be pointed out) is not a trivial and marginal aspect of life as a whole. If we are trying to understand what is making our friends unhappy, a detached approach will at some point not only distress them but completely block our effort to find out what is wrong. Or if we want to understand a profound play or novel, withdrawing our sympathy may make our attempt impossible.

But, much more widely, this happens also about a wide range of problems concerned with the motivation of other people whom we need to know about – outsiders and enemies as well as friends. Indeed it is true to some extent of all our attempts to grasp the kind of social phenomena which we mentioned earlier. Some degree of empathy or sympathy with the people involved is a vitally necessary cognitive tool for understanding any of them. We have to enter into their aims and intentions. This is why the kind of 'objectivity' which B.F. Skinner aimed at in psychology – the approach which abstracts from human subjectivity altogether, treating other people as though they were simply physical objects – is a cognitive dead-end.

THE PECULIAR STATUS OF PHYSICS

Physics, by contrast, stands, along with mathematics and logic, right at the other end of this spectrum. It is not just an immensely abstract enquiry but one which directs its abstraction specifically to shut out the peculiarities of personal experience. That is what makes it remote from ordinary thought. Its specialisation is entirely justified by its success in doing its own particular work. But the idea of using it as a place from which to explain the situation of consciousness, which stands right at the other end of the spectrum, is surely somewhat wild.

This is not to say that current attempts to alter the concepts of physics in a way which can make the existence of consciousness seem less paradoxical cannot be useful.[5] Whether or not they can work in scientific terms, they do help to undercut the crude, mechanistic idea of living organisms which at present, in many people's view, simply leaves no room for consciousness. But this is an indirect kind of facilitation. It works by altering biological concepts in a way that may allow them to mesh easily again with psychological ones. It is quite a different enterprise from fitting physics and subjective experience together directly while leaving out all the aspects of life that lie between them. A shotgun marriage with physics cannot be the right way to save the respectability of consciousness.

SUBJECTIVITY IS NOT SCANDALOUS

This shotgun marriage – this last kind of solution to the 'problem of consciousness' – is then neither workable nor necessary. It won't work because wide category-differences prevent the two from fitting together. And it is not necessary because consciousness already has a much more suitable partner. It exists in a workable symbiosis with its counterpart, which is something much wider than physics – namely the whole objective viewpoint, of which physics is only one specialised department. These two counterparts are combined in a wholly familiar manner over the whole range of our experience, and in a general way that range usually shows us how we had better connect them.

We are indeed often in difficulties about adjusting the details of that connection and these difficulties can sometimes be very grave. In particular, the value-questions about the relative importance of various elements in life, which I mentioned earlier, are vast and often leave us in doubt about how to balance our whole thought-structure. This is why serious literature and philosophy have plenty of work to do. But we do not suffer from the kind of general paralysis about these problems, the depth of mutual ignorance which would be expected if current materialist dogmas were correct. 'Consciousness' is not an isolated, peculiar phenomenon at the margin of the world, a surd item detached from the rational ways of thinking by which we make things intelligible. As Willis Harman put it in the first issue of the *Journal of Consciousness Studies*:

> When the conscious awareness of the scientist is conditioned by training to look outward only, the present form of science may seem to offer a reasonable world-view. But when consciousness turns back upon itself, and attention turns inward, not only is another realm of experience added to the picture, but a new order to external reality may be seen. The observer is changed in the process: never again can certainty be placed in the articulation of absolute laws that leave this factor of consciousness disregarded.
>
> ('The Scientific Explanation of Consciousness: Towards an Adequate Epistemology' by Willis Harman, *Journal of Consciousness Studies*, vol. 1, no. 1, 1994, p. 143)

For instance, as just mentioned, dentist and patient must do business together, and they can usually do so on the basis that they have some grasp of each other's viewpoints and of the social context in which they both operate. We all know that any explanation of matters connected with pain has to take both inner and outer aspects equally seriously. In medical contexts, therefore, the need to be objective about subjective experience is quite familiar to us. Yet as soon as we move away from a science (such as medicine) with a recognised physical basis, alarm and embarrassment descend. Strange as it may seem, many educated people today do find it hard to admit that the subjective viewpoint has an important influence on the world. That embarrassment is the point from which Searle starts his argument. He himself evidently shares it. As he puts it, this problem: 'has puzzled me for a long time: there are portions of the real world, objective facts in the world, that are facts only by human agreement. In a sense there are things that exist only because we believe them to exist.'

What is it that is puzzling about this? These alarming things are not hallucinatory or delusive objects, which it really would be odd to say 'exist because we believe them to exist'. They are the large-scale, well-recognised social institutions that we mentioned above such as laws and money, the kind of things that anthropologists and historians regularly observe and study. They are indeed arrangements produced by human agreement, but they are not in any way dependent on private fantasy. They are the kind of solid social fact against which (if we are not careful) we perpetually bang our shins in everyday life. Their reality becomes clear to us because we find that they can injure us. Like cities, they exist because people have made them and still 'believe in them' in the harmless sense that people think it worth while to maintain them. That is entirely different from saying that they exist because we believe them to exist.

It is surely odd that people should find a difficulty in counting things like this among the objective facts of the world. Has not something gone seriously wrong with the current notion of *existence* or *reality* when it is possible for such doubts to arise? But of course Searle is right in supposing that they do arise. A certain kind of naive, dogmatic materialism makes it look doubtful whether these things can properly be classed as wholly real and 'objective'. It makes these things – and the whole fact of consciousness – look like something that needs to be 'explained' in the special sense in which explaining things means justifying them, as in the phrase 'Explain your conduct!'.

What people want then is the kind of explanation that will show that – in spite of suspicious circumstances – these things do not constitute a scandal. The explainer must produce passports to prove that they are reputable parts of the world in good standing. Perhaps today this is indeed what is worrying some people most when they ask for an explanation of consciousness.

Along with the two aspirations mentioned earlier – the demand for a causal account and the enquiry about importance – it may well figure among the hopes with which people embark on 'the problem of consciousness'. Taking the status of this ideological passport-inspector for granted, they ask for an argument which will convince him that their everyday experience is indeed reputable and real.

SAVING THE APPEARANCES

This is a job that can be done. Searle accepts the demand and produces convincing passports. He retrieves the social facts that have slipped away through the cracks in theory. He shows how these institutions have developed in seamless continuity out of the increasing complexity of the acknowledged physical facts in the one world. He shows that no vitalism is involved. No illicit quasi-physical forces are needed. As he points out, the tendency to speak, and to crystallise accepted customs by ritual speaking, is a biologically-determined characteristic of our species, an emergent property which we have developed in the natural process of evolution. Humans need to map their world by imposing meaning communally on it in this way because of their highly complex social life. This technique for fine-tuning their existence is a further stage in the development begun by other social animals. For they too use rituals to establish their own customs, but rituals of a much simpler, less explicit, less flexible and less sophisticated kind.[6]

Social facts, then, are entirely continuous with biological facts. They are part of life. This shows how misleading the metaphor of 'social construction' is if it is taken in its strong sense as meaning that we are omnipotent in inventing social practices. Customs are not just an arbitrary product of our whim. Searle's central target is indeed the sense of artificiality and exaggerated power that this excessive constructivism suggests – the impression that it gives of human decisions taking place in a vacuum and dictating terms at will to the rest of the cosmos. He points out that human beings are not a separate entity detached from the natural world. Different academic disciplines, therefore, should not behave as if they each owned their own private universe. Physics, literary criticism, political theory, geology and ethics should all notice that they share a world. 'The traditional opposition that we tend to make between biology and culture is as misguided as the traditional opposition between body and mind . . . Culture is the form that biology takes' (p. 227).

This, the central theme of Searle's book, is surely dead right and very important. The Cartesian split is still rampant today. The abstractness and remoteness of much theorising in the humanities and social sciences echoes the reductive narrowness produced by pseudo-scientific materialism among people who hope they are following the banner of science. That narrowness is what led academics in a wide range of disciplines to book tickets on the behaviourist branch-line in the first place, and it is what still pulls them along it towards the futile terminus of full-scale metaphysical behaviourism. Though that terminus has in theory been renounced, it still attracts people who see that solutions half-way to it are unsatisfactory, but cannot yet nerve themselves to get off this track altogether.

'FUNDAMENTAL?'

Most of the time Searle shoots even-handedly at both targets. But he is still surprisingly impressed by the materialist story. He still uses a vital part of its language, the part that privileges explanation-by-matter over explanation-by-mind. Thus, to repeat his opening fanfare, having said that we live in exactly one world, he goes on to write: 'As far as we currently know, the most fundamental features of that world are as described by physics, chemistry and the other natural sciences.' And shortly after he asks, 'how does a mental reality . . . fit into a world consisting entirely of physical particles in fields of force?'.

But in what sense is that particular way of describing and analysing the world final or magisterial? How are the features of the world which physics describes more fundamental than its other features? Fundamental to what? There is something extremely mysterious about the traditional creed that is surely being invoked here, the creed which declares that (as Lewis Wolpert has lately put it): 'In a sense, all science aspires to be like physics, and physics aspires to be like mathematics.'[7] In short, abstraction is itself a prime aim. This would mean that all studies must try to get rid of their particular subject-matter so as to reach the condition of mathematics, which is universal because it has no subject-matter. They should try to stop talking about anything but universal forms.

Of course physics does have its own special importance. If the questions you are asking are already physical questions, then of course these are the appropriate terms to use. Modern physics is the study to which the pre-Socratic enquiries about the basic stuff of the world finally lead if you take them – as the Pre-Socratics themselves did not take them[8] – in a purely physical sense. When your question is 'what are the smaller physical elements of which this consists and the forces that govern them?' you quite properly end up doing physics, which is of course a necessary, noble and rightly influential enquiry. It is, so to speak, the North Pole at the end of all those signposts 'To The North' – the terminus if that is the way that you are going. But when you reach that pole you are at a remote level of abstraction from the world we usually live in and you have gone in a direction different from that of many equally important lines of enquiry – such as, for instance, enquiries about history or human motivation. Most of life calls on us to ask and answer questions different from this one. What is so special about the atomistic quest?

John Ziman, himself a physicist, puts the point well:

> The most astonishing achievements of science, intellectually and practically, have been in physics, which many people take to be the ideal of scientific knowledge. In fact, physics is a very special type of science, in which the subject-matter is deliberately chosen so as to be amenable to quantitative analysis ... Only quantities that can be represented numerically and transformed mathematically are permitted in the physical sciences. It is not simply good fortune that physics proves amenable to mathematical demonstration: it follows from careful choice of subject-matter ... Physics defines itself as *the science devoted to discovering, developing and refining those aspects of reality that are susceptible to mathematical analysis.*
>
> There is nothing fundamentally wrong with physics as such, but it is an inappropriate model for all consensible and potentially consensual knowledge.
>
> (From *Reliable Knowledge: An Exploration of the Grounds for Belief in Science*, Cambridge, Cambridge University Press, 1978, p. 28; author's emphasis)

THERE REALLY ARE MANY MAPS

I have suggested that the various maps of the world in our atlases do not need to compete for the prize of being called the most fundamental. It is just as true to say that 'the world consists' of lands and oceans, or of climatic zones, as that it consists of the territories of different nations, and so on. It is also as true to say any of these things as to say that it consists of physical particles in fields of force. *There is no bottom line.* Our choice of terms depends on the purpose for which we want to analyse the world at the time. The concession which Searle makes to dogmatic materialism in the wording of this passage (and throughout his book) is surely misleading and inconsistent with his powerful central point about the full reality of social phenomena. This concession damages his thesis in a way very like the one that occurs in the plaintive title of a book by Wolfgang Köhler, *The Place of Value in a World of Facts*. One might as well ask for the place of circles in a world of squares. If you always use graph paper, and take its squares to be the fundamental units of design, you will naturally be in trouble over circles.

PHYSICAL AND 'PHYSICAL'

One source of confusion here is the systematic ambiguity of the word 'physical'. That word is often used in an everyday sense that has nothing to do with the science of physics but is simply the opposite of 'mental'. Unconscious oscillation between these two senses is widespread in this whole controversy. Thus, Searle himself often uses the word simply to divide all facts into two classes, mental and physical. At other times, however, as in the phrase 'physical particles in fields of force', he uses it to mean 'recognised by the science of physics'. At one point (p. 196) he even describes our perception of colour as a 'subjective illusion', saying that 'colours as such are not a part of the external world' because physics does not recognise them. But – as we saw in our discussion of rainbows – it is not an illusion that letter-boxes in Ireland are green, along with the grass, while poppies and English letter-boxes are ineluctably red. It is a fact.

In the everyday sense where *physical* is opposed to *mental*, physical things are not the entities of physics. They are ordinary, macroscopic, unreconstructed items like apples and houses, things that are coloured, solid and eatable. That is why it sounds plausible that we live in a world composed of them. But we certainly don't live in a world composed of quarks and electrons, because physics simply has no place for us. Items such as *we* and *life* do not exist for it. They are outside its vocabulary. Physics follows Descartes' and Galileo's advice and refuses to talk about them. It deliberately describes an abstract, uninhabitable, life-free world.

When (therefore) Searle writes that 'we live in exactly one world' he cannot really be referring to the 'world' of physics. That world is not the whole caboodle but an idealised abstraction from it. Indeed, the word 'world' here is used almost as it is in the titles of journals such as *The Poultry World* to denote and rail off a special area of interest. This idealisation becomes increasingly obvious when we remember that physics itself constantly moves on, so that the 'ultimate particles' that it accepts at one time can easily be superseded ten years later.

Searle's trouble here is the same confusion that we saw in Richard Dawkins' remark, quoted in the Introduction, that 'Science is the only way we know to understand the real world'. Dawkins makes it plain that the kind of science he means here is essentially just particle physics. But the world which we need to understand – the world we actually live in – is in the first place a perceptual and social world, a turmoil of lights, colours and noises, love and hate, danger and hope, friends and enemies, plans and despairs. It has to be this kind of world because we are not pure observers but social animals of a particular species. (And if we were pure observers, is it so clear that we would necessarily find the search for a physical theory of everything more interesting than, say, questions about the nature of life? That, however, is not our problem.)

Of course we do also need to grasp the vast horizon that lies outside all this social and perceptual turmoil – the wider whole within which life and society arise. In Part 3 we will consider how best to do this. But we cannot deal effectively with that wider world by jumping off our shadows to go at once to the extreme of abstraction. Both the data which we use in theorising and the interpretations that we put on them arise within the range of our immediate concrete experience. They are profoundly shaped by the rest of our thought – by our weaknesses, our choices, our interests, our myths. If we do not bother to understand first what lies within that range of experience, we shall not reach much useful understanding of what lies outside it.

Neither houses nor quarks, then, are more real than mental items, as indeed Searle makes clear. Toothache is as real as teeth or electrons and debt is as real as the house that was bought with it. Everyday causality runs constantly and unhesitatingly across these borderlines all the time. The one world contains, without anomaly, all these kinds of entity – electrons and elections, apples and colours, toothaches and money and dreams, because it can legitimately be analysed in all these different ways. The various explanations that we need therefore involve, quite democratically, all the various kinds of thought that are needed to deal with them. As Rom Harré has pointed out,[9] theorists who have insisted on narrowing the language of causation in a way that seems to make these winding causal paths look impossible end up with a picture that is as unusable for science as it is for everyday life.

TERRITORIAL PROBLEMS

Thus our situation really is complex. But that complexity need not lead us into academic warfare about who owns the problem of consciousness. Such warfare is futile because this problem – or set of problems – is like the air, it encloses and concerns us all. Like many other topics it is complex enough to need intellectual co-operation between different kinds of thought, and that co-operation can always be organised.

Jeffrey Gray defined the question as 'how to get the conscious part of our existence, which is the most important part of our existence, into the scientific framework' and insisted that this must necessarily be 'a problem for the scientists to solve', certainly not one for philosophers.[10] And it is natural that scientists should want to keep this exciting topic to themselves. But unfortunately, even if they managed to keep outsiders away, they would still find that, in handling this matter, they themselves could not help doing philosophy. In order to make any progress at all, they would still be constantly forced to analyse concepts and to rebuild the background system of thought which connects them – a system which cannot help involving a wide range of disciplines.

There is no reason at all why scientists should not do philosophy and do it well. From Galileo and Darwin to Einstein and Heisenberg, many of them have done just that.[11] But philosophy is essentially about connections, so you cannot do it well if you insist on doing it in isolation. Protectionists who try to isolate the issue as Gray suggests are in fact almost bound to end up dealing with it badly, because – far from excluding philosophy – they are usually held captive by older philosophical views which they unconsciously take for granted, such as, in this case, those of Descartes.

The whole project of territorial rivalry about this topic is, then, an empty one. The world we live in is indeed one world and a mixed, unreconstructed, vernacular world at that, not a world purified and quarantined to serve a specialised 'science of consciousness'. The issues that arise when we try to think about consciousness are large. They affect the whole way in which we conceive our world. They need to be co-operatively handled because they present problems calling for every sort of explanation.

14

BEING SCIENTIFIC ABOUT OUR SELVES

'SCIENCE', OLD AND NEW

The main themes of this book involve two questions which are causing so much trouble at present that they are surely worrying most of us already. The first one is about self-knowledge, the other is about what it means to be *scientific*.

The word 'scientific' is troublesome because it now has two distinct meanings which contrast it with two distinct sets of opposites. On the one hand it can be a quite general word of praise, meaning simply *thorough* and *methodical* as opposed to casual, vague or amateurish. In that sense historians or linguists or logicians can be called scientific – or unscientific – just as properly as astronomers. On the other hand, the word can be a strictly factual one meaning 'concerned with the natural sciences' as opposed to other studies. In this sense (but not in the other) we can talk about *bad science*. In this sense, even a bad and casual book about astronomy is a scientific book, but a good and thorough book about history is *not scientific*. That, of course, is the familiar principle on which bookshops and libraries organise their shelving.

This is not a trivial ambiguity. It is part of a general confusion about the kind of praise that is conveyed by the notion of science, a confusion which is causing a lot of trouble. When the two meanings get mixed, it seems obvious by definition that the methods of the natural sciences are not just the best methods but the only ones that are intellectually respectable at all. They are therefore all that we need and can rightly be described as *omnicompetent* in the way that we have noted.

As we have seen, this idea that the methods of physical science are quite simply the only intellectual methods and should therefore be extended to cover every subject-matter, including our understanding of ourselves, was put forward early in the nineteenth century by Auguste Comte and others. It is still a powerful faith, devoutly preached by many people today. Only radical confusion about the meaning of the word 'scientific' makes it seem plausible. Before considering this confusion, however, let us think a little about the target area, the site to which it is now proposed to extend this empire – about my other troublesome concept, *self-knowledge*.

KNOWING OURSELVES

Self-knowledge is a notion that is often not much attended to when philosophers talk about 'the self' because the academic self commonly seems to be rather thin and easy to know. This may be because the discussion usually appears to be about other people's selves rather than one's own. Self-knowledge does of course raise puzzles (which do interest philosophers) about how subjects can somehow become their own objects – who is doing the knowing or controlling and who is being known or controlled? Such words make it clear how complex the human subject is, how many questions are involved in trying to understand it.

In everyday life, however, self-knowledge has a wider importance than this, one that crops up often in our personal affairs with a strong moral bearing. We often do find it hard to understand our own behaviour as well as other people's. But we usually accept that we still have to attempt this difficult kind of understanding. Total ignorance about our own motives, habits and capacities is not excusable. Knowledge about these things is not an optional subject like Russian or trigonometry which we can drop if we are not very good at it. Failure to know ourselves can be a serious moral fault. And one reason why it is a fault is that it blocks our understanding of other people. The sort of basic sympathy and empathy that we need in order to understand others does not work unless we are attentive to our own motives and reactions as well. Unless we ask critically how we ourselves are behaving to them, we can't hope to understand how they are behaving to us. So, surprisingly enough, in the enterprise of understanding other people, *cognitive success depends on moral attitude*. To get far in this study, you need fairness, honesty, maturity and indeed generosity. This means that there are facts which we cannot know unless we first get the values right. From the view of facts and values that has been widely accepted for much of the twentieth century, that is rather surprising.

This surprising fact also applies, however, to more general views about ourselves, views about the kind of entity that we and all other human beings are. What methods do we need for this study? The seventeenth-century poet Sir John Davies is instructive here:

> I know my soul hath power to know all things,
>   Yet she is blind and ignorant in all:
> I know I'm one of nature's little kings,
>   Yet to the least and vilest things am thrall.

I know my life's a pain and but a span;
  I know my sense is mocked in everything;
And, to conclude, I know myself a man –
  Which is a proud and yet a wretched thing.
                    (Closing lines of his poem 'Nosce Teipsum')

THE IMPORTANCE OF IMPORTANCE

Should we get sceptical about the way in which Davies talks about knowledge? Should we doubt whether he really does *know* these things and ask for further research? Ought we to say that this is only folk-psychology which must remain provisional till more work has been done about the cell-biology and the neurones? It is clear that this would merely be an evasion of his uncomfortable message. The facts that Davies mentions are well enough attested already to count as known because they form the habitual background to all our more detailed thinking. They are tacit knowledge. Further detailed evidence for them is not needed. The issues he raises do not concern the details of these facts but how we should respond to them. He is puzzled about the appropriate attitude to this curious self, an attitude which of course is not just a state of emotion but a considered, thoughtful response. He points out that there is good ground, not just for believing that human life is mixed and confusing in this way, but for taking a realistic, non-evasive attitude to its mixedness as a step to dealing with it better.

This involves a moral judgement about how we should proceed, a judgement about what is important, what we should attend to. That kind of judgement is always to be found at the root of metaphysics, including the apparently sceptical kinds of metaphysics (like materialism and determinism) which are sometimes inclined to deny that they are metaphysical at all. It is a judgement about what matters and what does not. These value-judgements are needed for selection. Without them, we could not form any general view about the human condition, because we could not decide which of a thousand patterns to pick out and study from the welter of experience.

Value-judgements about importance determine, among other things, what limits we set to the self itself, how far we think it extends and how sharply we separate it from what is around it. A self is not a given distinct object like a ball or a stone. For instance, the extreme individualistic model of selfhood – the social atomism which underlies social contract thinking – treats each self as independent, an object like a billiard-ball radically cut off from its fellows. But it does not do this on factual grounds. It is not a scientific discovery that selves are in fact shaped like billiard-balls. It arises chiefly out of moral indignation at the oppression which has often resulted from a more organic, connected, hierarchical view of social relations. Social atomism flows from deciding that the bad consequences of hierarchical systems are so important that the conceptual scheme underlying them must be ditched and replaced by a more separatist one.

At the other extreme, the Buddhist view that all separateness is an illusion – that individual selves are more or less arbitrary divisions across the continuum of life – also arises, not out of factual observation, but out of a sense of the harm we do by constantly cutting ourselves off from one another. The different value-judgements that underlie these metaphysical systems are essential to each of them, and besides deciding what counts as part of ourselves, these judgements also decide how we shall conceive the rest of the surrounding world. The independent self of the social contract lives in a world modelled on that of seventeenth-century cosmology. It is a solitary atom gyrating in a social void, a radically solitary rational entity moving among a crowd of others with whom it has no real connection. This is a world which does not easily find room for non-rational humans such as babies and a world which can scarcely accommodate non-human nature at all. The Buddhist self, by contrast, lives in a world without frontiers and must recognise a great range of other beings, human and otherwise, as literally continuous with itself.

These general ways of conceiving the world obviously make an enormous difference, not just to our notions about how we ought to act but also to our views about which facts we ought to attend to and what methods we should use in thinking about them. By affecting our selection of topics they alter our factual view of the world as well as our moral view about how we must deal with it. But neither of them is more scientific than the other, in the sense in which *scientific* is a term of praise. In that sense, either of them can be scientifically or unscientifically developed. We cannot use the idea of science as a criterion for judging between them.

WHAT KIND OF OBJECTIVITY?

This entanglement of fact with value is important because it shows how impossible it is to apply to these topics the kind of objectivity which we associate with physical science. That objectivity requires that all observers should stand at the same point of view and abstract from their individual differences. But this kind of abstraction simply cannot be used when we are talking about human affairs. There is no way in which we can collect facts about any significant aspect of human life without looking at them from some particular angle. We have to guide our selection by means of some value-judgement about what matters in it and what does not. And these judgements inevitably arise out of each enquirer's moral position. When such judgements raise difficulties, they need to be justified morally by explaining that position, not by ignoring it.

Social and psychological theorists who claim to be operating in a value-free vacuum outside morality are notoriously deceiving themselves. They simply haven't noticed their own biases. The behaviourist idea that, in order to be scientific, psychologists should study people *objectively* in the sense of ignoring their subjective point of view and treating them solely as physical objects was not value-free. It was not just a proposal for a new scientific method. It was a demand for a new and very peculiar moral attitude. In his last book, *Beyond Freedom and Dignity*, B.F. Skinner eventually made that moral attitude explicit, disclaiming all esteem for the active, creative aspect of human nature and flatly asserting the right of psychological experts to engineer it as they thought best. The exposure of these moral views probably played a large part in the subsequent discrediting of his methods.

Until then, however, the idea that a 'scientific' approach demanded this quite impossible abstraction from all views on what mattered in human life was widely accepted. It did not only allow psychologists to consider themselves scientists. It had the further and deeper advantage for them of exempting them, as professionals, from the painful efforts at sympathy and self-knowledge which normally form a crucial part of our attempts to understand other people. It allowed them a kind of detachment in which they could make a positive merit of not knowing or caring how the people whom they studied felt.

This affectation of detachment and ignorance about human experience was a good deal more extreme than what Comte and the other pioneers had in mind when they called for the founding of 'social sciences' in the nineteenth century. They simply wanted to use experimental and statistical methods drawn from physical science in the study of social affairs so as to make it more systematic. But they took it for granted that these methods could be added to existing ways of thought which were already useful there. For them, the term *scientific* mainly had the general meaning that it had held throughout the enlightenment. It meant thinking out problems afresh for oneself rather than relying on authority or tradition. They never envisaged throwing out all existing historical and philosophical methods and replacing them by ones drawn from physical science. Notoriously, Comte himself, when he talked of throwing out religion and meta-physics, only meant throwing out other people's religion and meta-physics and replacing them by better ones of his own invention.

These pioneers wanted to combine all useful methods. Thus the new social sciences started life as *sciences* in the old general sense – methodical forms of thought, parallel to history and logic, which would use whatever kinds of reasoning seemed helpful in handling their subject-matter. For instance, social scientists who discuss the scope of their own study, or its history, obviously cannot carry on that discussion by using the methods of physical science. What they are then doing is history or philosophy, whether they notice it or not. Thus Comte himself was not (in the modern sense) a scientist but a philosopher, though not a very good one. Similarly, people today who still echo Comte's philosophical manifestos are not doing physical science but are themselves philosophising, even if not very well.

POPPER'S GUILLOTINE AND THE DESTINY OF IDEOLOGIES

What finally broke up this hospitable, inclusive idea of 'science' was the steadily growing prestige of the physical sciences them-selves. That prestige led a growing number of prophets to jump on the bandwagon by claiming that their ideologies were *scientific* – not just in the old sense of being methodical but in the new one of being founded in some way on physical science. Herbert Spencer claimed scientific backing for social Darwinism. Marx also notoriously thought of his world-view as science-based and Engels strongly developed this claim, making the connection with physical science a central plank of later Marxist thought. Freud too relied heavily on claiming scientific status. All these prophets conceived of 'science' primarily in general terms as

Enlightenment thinking, the opposite of blind tradition and superstition. But they also believed, more specifically, that modern biology and physics directly underpinned their own social and psychological theories.

This proliferation of rival 'scientific' world-views was bound to bring its nemesis. Readers could not go on for ever adjusting themselves, like chameleons, to match all the different colours on this intellectual Turkish carpet and still call the result 'science'. Indeed it is rather surprising now to see how long the plurality was tolerated. Throughout the first half of the twentieth century intelligent people worked amazingly hard to combine obviously incompatible ideas together provided that they were all labelled 'scientific'.

After the Second World War, however, the guillotine came down. Karl Popper then pointed out that the Marxist and Freudian ideologies were not actually constructed by the methods of the physical sciences and could not therefore be described in modern terms as *scientific*. The unfortunate thing was that, at this point, he did not pause to ask what world-views of this kind were if they were not branches of science and what other standards they ought to be judged by. It should have been obvious that Marxism and Freudianism were indeed not primarily scientific theories but ideologies, comprehensive attitudes to life, with a strong moral component as well as their factual claims, and that they needed to be judged seriously by the standards appropriate to such general attitudes. Discussion of rival attitudes to life is not a vice nor a waste of time but an intellectual necessity, particularly in times of violent change. Popper's work, however, seemed to outlaw all such argument from the province of thought, ruling that, since it was not science, it was *metaphysics* – a word which he used vaguely and which many of his audience took to mean simply nonsense.

There followed a jubilant wave of crude scientism, not just in the sense that people put too high a value on science in comparison with other branches of learning or culture, but in the wider sense that they often forgot those other branches existed at all. This narrow vision now became explicit. Its prophets tend still to assume, as Atkins does in the recent article I have quoted, that the only available forms of thought other than 'science' are religion and parapsychology. They make no mention of history, law, language or logic and if they mention the social sciences at all they speak of them as dubious entities on the borders of science proper, becoming respectable only when they manage to imitate real science closely.[1] Thus the strange composite intellectual entity called science now turns out to be not just omnicompetent but unchallenged, the sole form of rational thinking.

Now of course this crude view is not universally held nor often defended explicitly today. Many scientists hate it. Atkins' open triumphalism is somewhat unusual. Yet, as I have suggested, as a myth, an imaginative pattern underlying more moderate thought, it still is very influential. Irrelevant notions about how to make thought 'hard' and scientific by imitating physical science still constantly distort the social sciences and many other areas of our thought, notably psychiatry. Though the enterprise of making all our thought on human affairs conform to physical patterns has never been at all successful, the idea that we must somehow do this impossible thing still haunts us. Many people find the prospect of abandoning that attempt unbearable.

For breaking out of this dilemma, three methods are currently being favoured. The first is to do research on some actual, existing, certified branch of physical science such as neurobiology or genetics, in the rather mysterious hope that, however irrelevant it may look at present, if it is pursued for long enough it will eventually throw some useful light on human affairs. This strategy requires a profound faith in reductivist methods which their record to date does not seem to justify. The second path is to re-describe human affairs themselves in a newly-invented set of terms which are reminiscent of those used in some physical science – terms which have a scientific look – hoping that this description can provide at last, a truly scientific explanation of human life. This is the path that has been followed by a great deal of research on artificial intelligence and, as we have seen, it is also the one taken by Richard Dawkins in his doctrine of memes.

I have suggested that that doctrine is an instructive example of the kind of wish-fulfilment produced by current confusions about what it means to be scientific. The reason why I spent some time examining memetics in Chapter 4 really is not just wanton destructiveness. As I mentioned there, I believe that unreal schemes like this are distracting us dangerously at a moment when psychology stands a chance of growing in much more useful and realistic directions. A century back, behaviour-ism was allowed to eliminate a whole host of ongoing psycho-logical enquiries on the quite mistaken ground that it was more 'scientific' than they were. As we now know, it wasn't. It had only managed to master the art of looking scientific by imposing a bogus simplicity: by imitating the externals of physical science. It offered a short cut past the really difficult issue of combining enquiries about the inner and outer aspects of our lives by ruling the inner one out of consideration altogether.

We do not need to make that mistake again. But if we continue to be hypnotised by an uncritical use of words like 'scientific' we are liable to do so. The dead hand of behaviourism will still control us, leading us always to prefer tidy thought-systems that have a vaguely 'scientific' appearance to ones that don't, regard-less of whether they are actually capable of telling us anything of the slightest interest about human life. Our natural resistance to self-knowledge – our chronic unwillingness to look into our own lives in the way that is necessary for real understanding of other people – always inclines us to accept schemes of this sort because they offer to distract us from these disturbing consider-ations. They make possible the neatly divided academic life whereby 'science' alone is pursued in the university, while all topics of real interest are left outside. This arrangement is, of course, always an option, even indeed a grant-attracting one. If we do not want to be stuck with it, there seems to be no substitute for self-knowledge.

THE SOURCES OF INDIVIDUALISM

Both these ways of trying to scientise our understanding of social affairs constitute distractions from our real problems rather than positive dangers. The third way, however, is influential and dangerous. That third approach is social atomism. It is the notion that, at the deepest level, human beings are not social animals at all but essentially solitary units, Democritean atoms spinning in a void. Though these atoms are hooked together externally for practical convenience, in their inner nature they are still isolated, still antiseptically separate.

As we have seen, this way of looking at things grew up in the seventeenth century for political reasons as a weapon in struggles for democracy and individual freedom. Its rhetoric went on developing as those struggles continued. Despite these political roots, however, it was always seen as a scientific doctrine, no doubt because of its likeness to physical atomism. As social mobility increased during the industrial revolution, this individualistic way of thinking steadily gained popularity and impetus, more especially in the United States where social mobility was at its highest, and it was at this point that the word 'individualism' was coined. De Tocqueville, whose view I noted in the Introduction to this book, introduced the term thus:

Of Individualism In Democratic Countries

I have shown how it is that, in ages of equality, every man seeks for his opinions within himself: I am now to show how it is that, in these same ages, all his feelings are turned towards himself alone. *Individualism* is a novel expression, to which a novel idea has given birth. Our fathers were only acquainted with *egoisme* (selfishness). Selfishness is a passionate and exaggerated love of self, which leads a man to connect everything with himself, and to prefer himself to everything in the world. Individualism is a mature and calm feeling, which disposes each member of the community to sever himself from the mass of his fellows, and to draw apart with his family and his friends: so that, after he has thus formed a little circle of his own, he willingly leaves society at large to itself. Selfishness originates in blind instinct: individualism proceeds from erroneous judgement more than from depraved feelings: it originates as much in deficiencies of mind as in perversity of heart.

Selfishness blights the germ of all virtue: individualism, at first, only blights the virtues of public life: but, in the long run, it attacks and destroys all others, and is at length absorbed in downright selfishness. Selfishness is a vice as old as the world, which does not belong to one form of society more than another: individualism is of democratic origin, and it threatens to spread in the same ratio as the equality of condition.

(A. de Tocqueville, *Democracy in America*, 1832,
Part 2, Book 2, Chapter 27)

He did not, of course, propose that democracy should therefore be abandoned. He was highly sympathetic to it and hoped that it would succeed. But he found the tendency alarming and he pointed out the need to do something about it. Other nineteenth-century thinkers, however, felt no such alarm. The social Darwinists endorsed the individualistic outlook unreservedly, ruling that competition between essentially separate individuals was in fact the inescapable basic law of life, and they claimed scientific warrant for this ruling by associating it with the name of Darwin. (This claim has been, incidentally, a main reason for creationist opposition to the theory of evolution in the United States, as Stephen Jay Gould demonstrates in a most interesting assessment of William Jennings Bryan's thinking.)[2] In our own time, sociobiology, which also claims to be scientific and Darwinian, has renewed this ruling about the fundamentalness of competition. Though sociobiologists do not spell out the practical social policies that went with social Darwinism, their crude rhetoric of selfishness unmistakably backs the metaphysic and the psychology that underlay those policies.

As we have seen, both these supposedly Darwinian doctrines are highly ideological. Both are visions developed in a somewhat hasty attempt to make sense of a socially mobile age. They are symptoms of the confusions produced by that age's fearful moral problems, not scientific resolutions of them. Indeed they are simplifications imposed in order to bypass those problems. The individualistic temper that they express and try to normalise is a by-product of our current way of life, not a human universal or a rational conclusion from scientific facts.

During the last half-century, the kind of uneasiness that de Tocqueville expressed about this whole development of individualism has been spreading in our culture. Despite a brief spasm of Thatcherism in the 1980s, most of us are no longer sure that there is no such thing as society. Instead, various attempts are being made to work out a more realistic, comprehensive view of the human social situation. Prominent among these is the concept of human rights, which we will discuss in the next chapter.

Part III

In What Kind of World?

15

WIDENING RESPONSIBILITIES

NEW WORDS, NEW MEANINGS, NEW NEEDS

It is striking how quickly the public has accepted the idea of human rights – only lately added to our moral vocabulary – as useful, indeed perhaps as indispensable, for talking about the world that we live in. Throughout the West, and to some extent throughout the whole international community, people understand roughly what is meant by violation of human rights. And they use that phrase at times to say things that they find very important. It is the sense of our times that, whatever doubts there may be about minor moral questions and whatever respect each culture may owe to its neighbours, there are real human wrongs. There are some things that should not be done to anybody anywhere. Against these things (people feel) every bystander can and ought to protest.

Academics are somewhat startled by this quick acceptance of the concept. They rightly point out uncertainties both about the central meaning of the term and about its borderlines. Yet in general the public is surely right to make use of the idea. This new conceptual tool is a powerful one and its power, like that of all such tools, is in some ways mysterious. Its full meaning is not easily explicated. It expresses rather more than we can yet put into our dictionaries. But this is the usual state of things where the world changes. New insights are responses to changes in the world and our vocabulary always has to evolve to meet such changes. New wine needs new bottles.

In this case, the change confronting us is a very general and radical one which is troubling us on many fronts. It is simply the immense enlargement of our moral scene, partly by sheer increase in the number of humans, partly by their greater mobility, partly by the wide diffusion of information about them, and partly by the dramatic increase in our own technological power. This power now enables us in the dominant nations to damage radically both the human and the non-human world that we live in, or to refrain from such damage.

The problem about distant humans is, I think, more closely linked both to the environmental problem and to the problem about the way in which we treat animals than has sometimes been noticed. (They are usually examined by different academic disciplines, which obscures their connection.) Technology has hugely multiplied both the range of matters that seem likely to concern us and our power of affecting those matters. And though that power often does not seem to be in our own individual hands, our civilisation as a whole clearly does bear some measure of responsibility for producing this whole situation. On this confusing scene, the idea of human rights is one of a number that have been evolved to help us select the most urgent points on which to concentrate our concern.

THE PROBLEM OF INCREDULITY

We find it hard, however, to use the idea of human rights effectively because of the sheer difficulty of believing in the whole expansion. Can it really be true (we ask) that we have duties to people so distant from us, people belonging to quite other communities? Can we, still more strangely, have duties to the non-human world? Both these things seem implausible because the changes that have taken place are simply too large and too rapid for us to take in imaginatively. We wander round the edges of these changes in bewilderment. Yet the changes are real. They demand some kind of adaptation from us, adaptation of a morality that was built for a quite different, much more manageable kind of world. We cannot go on acting as if we were still in that simpler world. On that path, we shall find no way through.

We are not, of course, alone here. This is not the first time that people in our culture have faced such a radical conceptual emergency. During the decline of the Roman Empire, people lost a mighty framework which had seemed central to the whole meaning of their civilisation. When the Goths took Rome, St Augustine offered Christians the idea of the city of God as a better replacement – an ideal city much less local, much less corrupted, but of course also more demanding. Other empires and cities too have fallen. People have repeatedly had to reshape their moral horizons. All such blows produce their own trauma and bewilderment, which has somehow to be met on each occasion by new moral thinking.

THE FLIGHT TO MORAL MINIMALISM

What form, then, should that thinking now take? One possible way of meeting the crisis is to widen the scope of morality, as St Augustine did and as the insistence on human rights does now. But another, which is certainly just as natural, is to narrow that scope. When changes remove social limits, exposing people to very wide demands, theorists may well restore those limits by contracting the social circle, by retreating into some kind of moral minimalism.

Since the Renaissance, this kind of contraction has in any case been happening in political philosophy in the West. Political thinkers of the Enlightenment systematically shrank morality by making it essentially a civic affair – a matter of mutual bargaining between prudent citizens within a limited society. Contract thinking sought to abolish the idea of duties towards anyone or anything outside that society. Of course the more subtle contract theorists, such as Kant and John Rawls, have not treated these duties simply as flowing from self-interest, as Hobbes did. But the original point of the model was to limit the scope of the duties within a definite society, not to enlarge that scope.

That limitation had originally a most respectable aim. It was meant to debunk supposed duties towards the supernatural because those duties had been used to justify fearful religious wars and oppressions. The real target of contract thinking was a distorted notion of duties towards God, and towards earthly rulers who claimed to be God's regents. But this move had an unintended side-effect. It now makes it quite hard for us to make sense of our responsibility towards humans outside our own society, and almost impossible to explain our responsibilities towards non-human nature.

THE WIDER HORIZON

Contract thinking, along with utilitarianism, created a bog in which philosophers who try to do environmental ethics are now busily floundering, alongside the similar bog that traps theorists of human rights.[1] Our habitual, official legalistic-cum-philosophic language does not have suitable terms for either of these things. This minimalism has not, however, formed the whole of our Enlightenment tradition on the matter. There was always another, contrasting strand of thought which resisted it and which has in fact generated current ideas of human rights. This was an expansive, hospitable, universalising, humanitarian movement which worked to counter the narrow legalism of contract thinking. Kant always combined both strands in a fertile dialectic, which is what makes him still one of our most useful philosophical ancestors.

That humanitarian movement's first great project was the eighteenth century's campaign for the Rights of Man. Campaigners for this extension attacked the meanness of limiting the idea of 'society' to certain privileged political units. They wanted to widen our duties so as to take in all the people excluded from those units, including foreigners and slaves. The language in which they expressed this proposal was, however, traditional. They used the existing idea of society though they widened its scope. They envisaged a super-society in which all rational beings (i.e. all humans, or at least all men) were citizens of the world, which thus became a kind of super-city-state. Kant expressed the point by talking of a kingdom of ends[2] – a realm of rational beings who, just because they were rational, existed as ends in themselves, not as mere means to the ends of others.

These reformers did not have to abandon the current political language because the idea of a wider city had already been pioneered by Augustine's talk of the city of God and before that by the Stoics, who also spoke of a world-city, Cosmopolis. But the attempt to invoke this vast image outside a religious context conflicted with the pragmatic, reductive thrust of contract thinking, which insisted on pointing out that all states have limits and that all obligations must lie inside what is practical.

THE ENLIGHTENMENT'S DIALECTIC

These two elements in the Enlightenment vision clashed at once in Rousseau's thinking and they have continued to clash ever since. For, in spite of much cheerful boasting by Nietzsche and the post-moderns, the Enlightenment is not something safely tucked away in the past, something obsolete that we can patronise. It is where we still live. Post-modern insights are themselves a recognisable part of it. That movement still sets our current moral scene simply because we have not yet managed to resolve the deep clashes that arise between the various elements that were jammed together in its message – clashes between order and freedom, between feeling and reason, between humanitarianism and the rule of law. Those clashes are the real source of our present problems.

Of course these clashes have not been merely destructive. They have provided a dialectic which makes possible the kind of development that any morality needs. Realistic reductivism has repeatedly done a real service to Utopian visionaries by forcing them to deal with nitty-gritty problems of detail. The effective reformers have been the people who have managed to operate on both sides of this divide, combining the two elements in their own thinking rather than fighting battles over them with outside enemies. They have been people who somehow contrived to fuse moral realism in this ordinary sense – not in the technical one which we will discuss presently – with bold aspiration, making practical thinking the instrument of their ideals rather than wasting effort in friction between the two. They have managed to interpret their long-term, demanding principles in terms of something immediately practicable without losing sight of the longer aim. This kind of double vision is in fact the central thing that can make any kind of moral progress possible.

That feat is, however, extremely hard. The two strands of thought always tend to keep coming apart. Reforming movements, as they grow, perpetually find themselves, to their members' distress, splitting into extreme and moderate factions. Idealists dismiss those who concentrate on small, relatively practicable measures as mere compromisers while people who feel themselves to be practical accuse idealists of utopian fantasy and humbug. Over our present topic – which is the scope of possible duties – this drama is usually still played out in the Enlightenment's terms as a conflict between legalistic reductivism and human sympathy, a conflict which is easily generalised into a wider war between reason and feeling.

But these seductive formulations are always misleading. Thought and feeling, law and sympathy are not – as Enlightenment thinkers too often supposed – separate imperialistic powers fighting a battle to take over the command of a country called Morality. They are all complementary elements within it. The notion that they can be treated as belligerents or duellists is, however, still prevalent and it is the source of our debates about human rights, as it is of those about environmental ethics and about duties to animals. There, too, the argument surges mainly round attempts to find, somewhere within the conventional vocabulary, language which will give these matters the status that they now seem to need. And there, too, these attempts continually run up against the limits set by the tradition.

The notion that the environment in general – nature as a whole – has value in itself, a notion which is expressed in views such as 'deep ecology', is blocked by the more reductive idea that nature cannot possibly matter except in so far as it contributes to human welfare. About animals, things are still more confused. The idea of dismissing them as mere disposable instruments now strikes many of us as immoral and repulsive. Yet the rationalist half of the tradition is deeply committed to claiming that only rational beings of a strictly human kind can have the kind of value or importance that would bring them within the scope of morality at all.

The persistent power of this view in official quarters can be seen in a recent move by the Charity Commissioners in Britain, who warned the Royal Society for the Prevention of Cruelty to Animals that it might lose its status as a charity if it went on campaigning against forms of animal-abuse which were advantageous to humans. In a letter to *The Guardian* the Chief Charity Commissioner himself explained that campaigns for animal protection could only count as charitable in so far as they were aimed at 'raising public morality by repressing brutality and cruelty *and thereby elevating the human race by stimulating compassion*'.[3]

This principle – evidently not just his own opinion but the official remit of the Commission – rules that charities must confine themselves to doing good to the human race. They cannot attend to non-human damage or suffering. On this principle, there is nothing at all wrong with cruel practices towards animals in themselves. They only become wrong if they make people less compassionate towards other humans. This is a causal link which would be uncommonly hard to establish. After all, might not some ill-tempered people treat their families even worse, not better, if they were forbidden to take out their bad moods on the dog? But this does not seem to settle the question whether they ought to be allowed to do so ... The 'human advantage' which is held to justify these practices is evidently rather widely interpreted by the Commission to include simple profit, since the examples under discussion included, not just medical research but the trade in the live export of calves from Britain, as well as hunting.

This bizarre piece of eighteenth-century rationalism is surely out of tune with our current moral insights. In fact, it has probably only managed to hold its place for so long because of the convenient twist which has so far allowed campaigning on behalf of animals to proceed under indirect licence, nominally as a way of improving the human character. If the literal meaning of this remit continues to be pressed, public opinion will probably force it to be altered. On this point, as well as about the wider environment, the reductive, rationalist tradition no longer represents current moral thinking. As quite often happens, the public is ahead of the academics morally and the conceptual scheme needs to be reshaped.

THE POINT OF EXCLUSIVE HUMANISM

This dominant emphasis on human claims was not originally meant to work as a barrier against concern for other earthly beings. As just mentioned, it simply reflected a political campaign against the use of religion to justify exploitation. It put forward a narrow, contractual view of political obligation as a defence against the abuse of religiously-motivated loyalty by self-interested rulers. And it was, of course, invented at a time when the facts about our planet were quite different from what they now are. The possibility of real environmental disaster had not then come on the horizon at all. Its appearance there in the last half-century is surely one of the greatest changes that has ever happened to the human race. It would be surprising if it did not demand an answering change in everybody's conceptual scheme.

In the earlier history of the West, however, the narrow, contractual view of political obligation had become rather strongly entrenched. More generally, it had given rise to a tendentious, reductive notion of rationality itself as essentially the calculation of self-interest. This notion is still perpetuated in the language of economics and it surfaces whenever altruistic claims are brought forward in public debate. In particular, it furnishes a background which can make it seem flatly impossible for rational people to extend the notion of rights to remote humans or to animals, or to be directly concerned for the environment.

IS IT JUST A VERBAL QUESTION?

There is a real difficulty here because the actual word 'rights' does have strong connections with the lawcourts. It is by its nature forensic. It easily looks competitive and litigious. In law, 'rights' are always rights held by one person against another or others. They can also only be held by beings who are able in principle to appear and argue their case as litigants in court. Though these requirements are actually fudged for infants and other incapacitated humans by allowing others to represent them, many people still think that this makes it impossible that apes or elephants should have rights. Right-bearers, they feel, need to be standard people directly involved in the political process even if they cannot actually stand up and speak in court. They should be fellow citizens of some kind, a requirement which even the best elephants cannot meet. Still less are the Antarctic and the rain forest going to appear as litigants. And in the main tradition, each country has also viewed the inhabitants of foreign countries as standing outside the rights conferred by its own legal system.

The important question here is whether this is just a verbal restriction on the use of the word *right* or a substantial point about what we ought to do, a real limitation on the realm of what can rightly concern us. Over animals, it is often possible simply to bypass the word *rights* and to talk instead of our duties or responsibilities towards them. This use of the word *duty*, however, has also been blacklisted by some moral theorists, who have insisted that 'rights and duties are correlative' – not just in the weak sense that if someone has a right someone else must have a duty to meet it, but in the much stronger sense that no one can have a duty to anyone who cannot owe them duty in return.[4]

This idea is not very plausible in the case of incapable humans. Indeed it surely shows up the unreality of this whole contractual, rationalist approach. *Responsibility* however still remains as a less objectionable word for the claims that animals have on us. It is also quite a convenient one for cases like the Antarctic and the rain forest, for which the word *right* is in any case less often invoked than *value*.[5] With this kind of rephrasing, many people find it possible to agree that there are real claims here – ones on which we must actually act – even though there is not, in the legal sense, a right.

16

THE PROBLEM OF HUMBUG

WORDS ARE NOT JUST AIR

Can we use the sort of circumlocution that we have just been discussing for all these problems? Can we simply avoid using the word right? In the case of distant humans, it seems that we cannot. People invoke the word right here for the same reason that led eighteenth-century reformers to talk about the rights of man – namely, its implacable force. Words like duty and responsibility tend to seem weaker in that they can be seen as reporting primarily facts about the person who is under obligation. You or I have a duty – perhaps we will perform it, perhaps not; who knows? This may look like our own affair, private between us and our own consciences. And in many large-scale cases it is not at all clear just who the duty or responsibility belongs to. By contrast, to talk of rights is to talk directly about the people who need relief. It aims to lay a burden publicly on anyone who stands in the way of relieving them – a burden which cannot be dodged by passing the can. The quasi-legal language invokes the broad impersonality of the law. It makes it much harder to say 'this is none of my business'.

So the inducement to use the word rights for these very wide, extra-legal purposes is strong, even though constant difficulties arise about specifying the details of its meaning when we have done so. The down-side of that advantage is, of course, the drawback that always goes with employing stronger language. People who think this language inappropriate may simply label it as empty rhetoric and dismiss one's whole claim without even asking themselves about the substantial moral issue behind it – as if the mistakenness of the wording settled the matter. This has indeed happened to a considerable extent about 'animal rights', especially in Britain.

About human rights it is not happening at present half so much, at least at the public level. Among academics, however, this confusion between the verbal and the substantial level is surely rampant. There are some philosophers who seem to view the whole notion of 'moral rights' – i.e. non-legal ones – simply as a weed much like ground-elder, a nuisance that perpetually recurs because of public incompetence and must just be uprooted every time it does so.

In such cases, however, it is surely better to assume that people are trying to say something genuinely important, however faulty their language may be, and to work to improve that language. We should try to take the pressure off the hotly-debated word by looking around for near-synonyms by which to elucidate it rather than flatly taking sides for or against it. I have regularly taken this line myself about the rights of animals.[1] In the United States, however, the Constitution has given the word rights a quite special importance which leads some people to think it indispensable for establishing the seriousness of any moral question. That is the view of Tom Regan, which he has argued in a number of books, notably in *The Case For Animal Rights*.[2]

WHAT LIES BEHIND THE LAW?

What, then, should we do about the case of humans? As just noted, when legal words like *right* are used outside the context of the lawcourts the point of using them lies in their somewhat paradoxical force. The idea is to extend the authority of law, for overwhelming reasons, to areas that no law can currently reach. On a good day, this language can work to bring these more distant cases fully and flatly into the same light as the ones that the laws of our own society literally require. The assumption behind the move is that we can appeal from law itself to some set of underlying principles that determine what is right and wrong, principles which always extend beyond existing law and are the source of that law's moral authority.

This assumption has often been expressed metaphorically by talk of a deeper, underlying law – natural law or the law of God or the moral law that Kant spoke of. These metaphors have considerable force but they also have well-known weaknesses. A great deal of ink has been spilt in discussing their faults and in trying to find ways of doing their work better. Instead of getting involved in these disputes here, I want to suggest instead a rather simple approach to them which may be helpful for our present subject.

THE FEAR OF MORAL EXCESS

I suggest that, in practice, we all accept the idea of trying to reach consensus about deeper principles and using them to correct existing laws and customs. Everyone who takes moral questions seriously at all does this and thinks it right to do it. No reasonable person is really an extreme relativist of the kind that might suppose everything currently allowed to be equally right or 'valid'. Almost everyone at times finds existing law and custom utterly inadequate and suggests standards by which to change them.

The general standing of this kind of critical thinking is not really in question. Nor ought the difficulty of clearly formulating these deeper principles to stop us from keeping on trying to do it, any more than similar difficulty ought to stop us on any other troublesome intellectual quest. What does come in question – what does raise objections to the whole search for these deeper principles – is the apparently excessive nature of the demands they make once they are found.

These wider principles tend to ask for much more than the detailed laws that they claim to supplant. They tell us, for instance, to love our neighbours as ourselves, or to treat other people always as ends and not as means only, or to secure the human rights of people who are entirely out of our reach. The charge against people who make proposals like these is not just one of obscurity. It is a charge of unreality, of humbug, of indulging in fantasy. This is itself a moral accusation, which is why it is often somewhat virulently brought.

The issue of humbug needs close attention here because it raises big issues about the general relation between ideals and practice. On this topic a somewhat obvious point needs to be made. The mere fact that we recognise the importance of ideals that are better than our existing practice does not in itself mean that we are hypocrites or humbugs. It is normal. The whole point of having ideals at all is to criticise current practice. If there were a society whose ideals were no more than a description of its existing behaviour it would be almost inconceivably inert, far more so than any culture that we have actually heard of. Even systematically static cultures have to make an ideal of their static-ness, actively resisting change, and to select issues that demand attention. A set of people who did not employ ideals even to this extent would have no mainspring for further action at all.

Ideals, then, do not become inert or unreal merely because they are far above us or ahead of us and are not likely to be reached in our time. That is their nature. They exercise their pull all the time by indicating a direction. Any existing situation, however bad, is still always the result of a struggle between prevailing ideals and the forces that resist them. The idea that it would be realistic to ignore or discount the force of those ideals is not sensible. Indeed it is itself a fantasy. It is an attempt to pretend that human life is far simpler than it is.[3]

It is quite true that this gap between ideals and practice does easily give rise to real humbug, and that it does so more readily when the ideals are specially demanding. Hypocrisy is, as they say, the tribute that vice pays to virtue. We are therefore very prone to suspect that this is happening. It is also true that newly proposed ideals shock us particularly easily in this way. (Once they have grown more familiar, we get used to the discrepancy.) That suspicion works at present to make many people somewhat incredulous both about claims for human rights and about calls for environmental action.

It is surely important, however, not to attach too much importance to this chronic suspicion of humbug, because it so obviously arises about other kinds of claim too. Indeed it attends ethics generally. It tends to arise whenever we seem to be called on to do something new and inconvenient. Sometimes the suspicion of unreality turns out to be justified, sometimes not. In itself, however, this kind of suspicion is no more than a sort of warning-signal telling us to ask ourselves whether these particular duties are real or illusory ones.

THE DIFFICULTY OF REAL REALISM

By contrast, more reductive, debunking models get an easy hold on our imaginations because they are so familiar. That fact tends to make them seem more realistic and, when wider claims are advanced against them, there is always an initial sense of unreality, perhaps of humbug. But there does not seem to be much real force in this impression. What passes for a realistic attitude in morals commonly means little more than addiction to current habits. That addiction is just as resistant to long-term prudence as it is to altruism. When new dangers appear, even very serious dangers (as has happened over the destruction of the environment) the *soidisant* realist usually has great difficulty in believing that they may actually be real. Genuinely enlightened self-interest is rare and hard to come by – which is one reason why egoism cannot be relied on alone as the basis of morality.

This kind of 'realist' habitually dismisses all long-term views and unfamiliar projects on principle as utopian and is (in particular) systematically convinced at present that all environmental protesters must have got their facts wrong. That kind of general, knee-jerk habit of suspicion is not, then, particularly important. Force of habit is extraordinarily strong, much stronger than the urge for self-preservation. It still produces this kind of incredulity even when the path that we are used to following plainly leads us towards the edge of a cliff.

WHAT IS A REALIST?

So far, I have (as I explained) been using the word *realist* in its general, everyday sense. It seems necessary, however, also to consider the special, technical sense that it bears at present in the study of international relations. I have to confess that I did not know of this special sense before I began, fairly recently, to look into questions about human rights. Outside that sphere I knew of just two meanings for the word realism. I knew (first) the general moral one, which we have been considering so far, in which the word means honesty about unpleasant facts, as opposed to fantasy, panglossism, sentimentality or humbug. I also knew the quite different scientific sense where *realism* means the belief that certain entities (such as quarks or electrons) are actual things in the world rather than just convenient concepts which can be used in calculation. In this sense the opposite of *realism* is *operationalism*.[4]

It is surprising, then, to discover that, in the field of inter-national relations, these two dissimilar senses have apparently been fused to yield the view that – to put it rather crudely – honesty calls on us to recognise the nation state as a specially real entity, a peculiarly hard fact in the world, a unit so uniquely solid and objective that it can fix the limits of our moral obliga-tions. In view of the changes that have continually taken place during history in the way the world is organised, this seems to be a remarkably arbitrary proposition. It is not obvious why the burden of proof should be put on anyone who proposes that people can have duties to other people outside their own nation state rather than on someone who does not. The charge of fantasy, sentimentality or humbug, which the use of words like realism always suggests, is not in itself an argument for this restriction. It is just an all-purpose psychological weapon available against unwelcome demands of every kind.

The unit that we call the nation state is, one would suppose, just a convenient division, in itself neither bad nor good. It is, as they say, a social construction, not only in the sense in which all our ideas are so, but in the much stronger sense in which our political and social arrangements are so and the sun and stars are not – a matter which we discussed in Chapters 5 and 12. We in the West have invented nation states during the last few centuries for whatever human purposes we have happened to have at the time. So it is up to us to attach to them today the kind of moral importance that we think they ought now to have.

Currently, financial practice and the general speeding-up of communication seem to be downgrading the role of the nation state relative to that of various other entities and forces. Notori-ously (for instance) the assets of some multinational firms exceed those of some small countries which are members of the United Nations, so they are likely to have greater power. There is also at present a considerable worldwide surge of various 'nationalist' movements committed to redrawing the boundaries of existing states in a way that will better suit their 'nations' in the sense of ethnic or cultural groups. For some time, too, an uncomfortable discrepancy has been obvious between nation states of varying sizes. Monaco, Norway and China are not really units of the same kind. In the face of all these changing factors, political theories that give the nation state a special, immutable moral importance surely need to defend their findings morally, by showing that this is the right way to treat it. There is no *a priori* reason why it should be seen as a general limit of duty.

We do meet here, of course, the problem of practicability which is set by all enlargements of duty. Even if our nation state is not the ultimate moral boundary, this cannot mean that we actually have a duty to help everybody outside it because our powers to do so are always limited – as, indeed, is also true of helping people inside it. That perfectly sound point is what political 'realists' often have in mind, and so far as that is what they are saying, their claim to realism is justified. But the question how far our powers actually extend is not at all a simple one. Astonishing feats have often been performed by people who simply decided to view something as possible which others were saying was not so, as for instance the Athenians did when they decided to oppose Persia at the battles of Marathon and Salamis or as Florence Nightingale did when she invented nursing. We can often do enormously more than we are inclined to claim. And when we belong to powerful nations, our public opinion can undoubtedly sometimes influence the behaviour of foreign governments.

CONCLUSION

Is there any reason, then, why it should not make good sense to say that we can have a duty to help their ill-treated subjects to the limit of our powers? I should perhaps reminisce a little here to explain why I find it so hard to see how anyone can doubt this. I myself began to read the newspapers around 1930, at the time when the Nazis were rising to power. Anyone who attended to this process saw, I think, that it was not just a local affair for the Germans but was the business of everyone in Europe. It altered the colour of the sky for everyone. And this was not just because the threat of another war might damage the local interests of people elsewhere. It was because of the specific moral moves that the Nazis were making, moves which mattered to everybody.

The unbridled nationalism, the propaganda for racism, the claim to justify brutal methods of repression, the habitual murder of opponents, the general cultivation of hatred were the direct concern of us all. These things did not strike us as merely the unavoidable eccentricities of bizarre foreigners but either as evils which we ought somehow to resist or – in the case of those who supported Nazism – as a creed to be welcomed. Nazi Germany, like Stalinist Russia, fell within our own moral universe. Hard though it might be to do so, we thought we ought to try and help the victims of both. And in the early days of both regimes we in the West could indeed have put on pressure to do so.

Thus, those of us who then noticed what was going on received at once the culture shock which only reached a much wider public after the war, the shock which later gave rise to the general endorsement of human rights. Serious journalists were already telling us, at that time, quite enough about the concentration camps, about the policy of holocaust and about Stalin's treason-trials to make us see that, whether we liked it or not, we were living in what the Stoics called Cosmopolis and the people to whom these things were happening were our fellow citizens in it.

There was no way of isolating our country morally from them, no barrier that could make these things cease to be our business. Nor does anything that has happened since then prove that view to have been mistaken. Indeed, since then the interdependence of distant nations has been steadily increasing. It surely seems quite as evident now that abuses in Nigeria or Tibet are our business as it then did that those in Russia or Germany were so. The idea of 'human rights' seems now to have become an essential tool in making this point clear and we must deal constructively with whatever difficulties may arise in a way that ensures we can continue to use it.

17

INDIVIDUALISM AND THE CONCEPT OF GAIA

WORLDS AND WORLD-PICTURES

This notion that our responsibilities do not end at national frontiers – that we owe some real duty to other humans – is not, of course, a new one in our culture. It has quite strong traditional roots, which have always warred against the narrower contractual view. The idea that we might also owe duties to the non-human world is, however, much more shocking. The contractual model of rationality excludes that idea and our tradition has taken some pains to stigmatise it as sentimental, pagan and anti-human. And until lately, prudence did not seem to call for this kind of consideration either because the natural resources available to us were seen as literally infinite. As the Soviet historian Pokrovskiy put it in 1931: 'It is easy to foresee that, in the future, when science and technique have attained to a perfection which we are as yet unable to visualise, nature will become soft wax in man's hands, which he will be able to cast into whatever form he chooses' (M.N. Pokrovskiy, *A Brief History of Russia*, 10th edition, 1931, translated and quoted in I.M. Matley, 'The Marxist Approach to the Geographical Environment', *Association of American Geographers' Annals* 56, 1966, p. 101).

This kind of confidence, generated by the industrial revolution, seemed for a long time to be a mere dictate of rationality, a simple correction of the earlier awe and respect for nature which now appeared primitive and superstitious. That is why we now find it so hard to take in the evidence that there was an enormous factual mistake here. For three centuries we had been encouraged to consider the earth simply as an inert and bottomless larder stocked for our needs. To be forced to suspect now that it is instead a living system, a system on whose continued activity we are dependent, a system which is vulnerable and capable of failing, is extremely unnerving.

Yet the damage already done undoubtedly shows that this is so.

How can we adjust to this change? As I have suggested throughout this book, in conceptual emergencies like this what we have to attend to is the nature of our imaginative visions – the world-pictures by which we live. In the vision belonging to the contractual tradition, the natural world existed only as a static background. It was imagined simply as a convenient stage to accommodate the human drama. That vision radically obscured the fact that we are ourselves an organic part of this world, that we are not detached observers but living creatures continuous with all other such creatures and constantly acting upon them. It blinded us to the thought that we might be responsible for the effect of these actions.

In order now to shake the grip of that powerful vision what we need, as usual, is a different one that will shift it. We need a more realistic picture of the way the earth works, a picture which will correct the delusive idea that we are either engineers who can redesign our planet or chance passengers who can detach themselves from it when they please. I think that we need, in fact, the idea of Gaia.

WHY GAIAN THINKING IS NOT A LUXURY

The idea of Gaia – of life on earth as a self-sustaining natural system – is not a gratuitous, semi-mystical fantasy. It is a useful idea, a cure for distortions that spoil our current world-view. Its most obvious use is, of course, in suggesting practical solutions to environmental problems. But, more widely, as I am suggesting, it also attacks deeper tangles which now block our thinking. We are bewildered by the thought that we might have a duty to something so clearly non-human. But we are also puzzled about how we should view ourselves. Current ways of thought still tend to trap us in the narrow, atomistic, seventeenth-century image of social life which grounds today's crude and arid individualism. A more realistic view of the earth can, I think, give us a more realistic view of ourselves as its inhabitants. Indeed we are already moving in this direction. But we need to do it much more clear-headedly.

This issue is not just psychological: it affects the whole of life. Our ideas about our place in the world pervade all our thought, along with the imagery that expresses them, constantly determining what questions we ask and what answers can seem possible. They enter into all our decision-making. Twists in those imaginative areas surely account for the curious difficulty that we still have in taking the environmental crisis quite seriously – in grasping the place that it ought to have in our scheme of priorities.

WHAT, THEN IS THE THEORY?

The current Gaian thinking that I believe can help here is a new scientific development of an old concept. The imaginative vision behind it – the idea of our planet as in some sense a single organism – is, of course, very old. Plato called the earth 'a single great living creature' and this is language that people in many cultures would find natural.[1] Our own culture, however, shut out this notion for a long time from serious thought. Orthodox Christian doctrine damned it as involving pagan nature-worship. And modern scientists, for their part, were for a long time so exclusively devoted to atomistic and reductive explanations that they too rejected this reference to a wider whole. Indeed, during much of the twentieth century the very word 'holistic' has served in some scientific circles simply as a term of abuse.

Recently, however, scientists have been becoming somewhat less wedded to this odd one-sided reductive ideology – less sure that nothing is really science except particle physics. The environmental crisis has helped this shift by making clear the huge importance of ecology, which always refers outwards from particulars to larger wholes. In that changed context, solid scientific reasons have emerged for thinking that the notion of our biosphere as a self-maintaining system – analogous in some sense to individual organisms – is not just a useful but actually a scientifically necessary one.

It is not surprising that an idea should combine scientific and moral importance in this way. As we have seen, science is not just an inert store of neutral facts. Its facts are always organised according to patterns which are drawn from ordinary thinking in the first place (where else, after all, could they come from?) and which often rebound in a changed form to affect it profoundly in their turn. These strong pieces of imaginative equipment need to be understood and criticised in both their aspects. We should not slide into accepting their apparent moral implications merely because they are presented as part of science.

The two-way influence of imagery is shown impressively by

the powerful machine image which was central both to the Newtonian view of the cosmos and to the Enlightenment's notion of determinism. As Karl Popper put it, 'Physical determinism . . . was a daydream of omniscience which seemed to become more real with every advance of physics until it became an apparently inescapable nightmare'.[2] The machine-imagery had taken charge of the thought. Another striking example today is the neo-Darwinist picture − now extremely influential − of evolution as essentially a simple projection of the money market. Here the noisy rhetoric of *selfishness, spite, exploitation, manipulation, investment, insurance* and *war-games* easily persuades people that this new form of Victorian social-atomist ideology must be true because it has the support of science. By using a different imagery and a different basic pattern, Gaian thinking tends to correct this outdated bias. It does not reject the central scientific message of neo-Darwinism about the importance of natural selection. It simply points out that it is not the whole story. Making this clear is, indeed, one of its more obvious advantages.

PLANETARY CONSIDERATIONS

I have been suggesting that this way of thinking has implications far beyond science. But the scientific case for it must be sketched first, however inadequately, so as to make clear what the term 'Gaia' actually means today. I shall summarise it here and shall then only return to it briefly at the very end of the book in considering its practical applications for immediate environmental problems. At present we shall be occupied with the idea itself and its place in our life.

The idea first arose out of considerations about the difference between the earth and its sibling planets. James Lovelock was employed by NASA in the early 1960s, designing sensitive instruments that would analyse the surfaces and atmospheres of other planets. But it seemed to him that the experiments proposed for detecting life on other planets were too closely bound to expecting particular features similar to life on earth. Was a wider strategy possible? Perhaps, he thought,

> the most certain way to detect life on planets was to analyse their atmospheres ... life on a planet would be obliged to use the atmosphere and oceans as conveyors of raw materials and depositories for the products of its metabolism. This would change the chemical composition of the atmosphere so as to render it recognisably different from the atmosphere of a lifeless planet.
>
> (James Lovelock, *The Ages of Gaia*, Oxford,
> Oxford University Press, 1988, p. 5)

He therefore compared the atmospheres of Mars and Venus with that of the earth and found indeed a startling difference. By this test Mars and Venus appeared, in a simple sense, static and dead. They

> had atmospheres close to equilibrium, like exhaust gases, and both were dominated by the generally unreactive gas carbon dioxide. [By contrast] the earth, the only planet that we know to bear life, is in a deep state of disequilibrium ... Earth's atmosphere is like a dilute form of the energy-rich mixture that enters the intake manifold of a car before combustion: hydrocarbons and oxygen mixed ... An awesome thought came to me. The earth's atmosphere was an extraordinary and unstable mixture of gases, yet I knew that it was constant in composition over long periods of time. *Could it be that life on earth not only made the atmosphere but also regulated it* – keeping it at a constant composition and at a level favourable for organisms?
>
> (Lovelock, *Gaia, The Practical Science of Planetary Medicine*,
> London, Gaia Books Ltd., 1991, pp. 21–2; emphasis mine)

Checking what might follow from this, Lovelock found that there is indeed a whole range of mechanisms by which the presence of life seems, from its first appearance on the earth, to have deeply influenced the atmosphere in a way that made its own continuance possible when it otherwise would not have been.

The scale on which this happens is hard to grasp. We need only consider here one simple and dramatic element in it – the carbon cycle. The carbon which living things use to form their bodies mostly comes, directly or indirectly, from carbon dioxide – the somewhat inert gas which, on the other planets, acts as a full-stop to atmospheric reactions. Life is therefore always withdrawing this gas from the atmosphere and two statistics may convey something of the scale on which it does it. First, if you stand on the cliffs of Dover, you have beneath you *hundreds of metres of chalk* – tiny shells left by the creatures of an ancient ocean. These shells are made of calcium carbonate, using carbon that mostly came from the air via the weathering of rocks – the reaction of carbon dioxide with basaltic rock dissolved by rain.

This process of rock-weathering can itself take place without life. But when life is present – when organisms are working on the rock and the earth that surrounds it – it takes place *one thousand times faster* than it would on sterile rock.[3] Coal and oil, similarly, are storehouses of carbon withdrawn from the air. All this carbon will go back into circulation one day, but meanwhile it is locked away, leaving the breathable air that we know, air that makes possible the manifold operations of life. Similar life-driven cycles can be traced for other essential substances such as oxygen, nitrogen, sulphur and that more familiar precious thing, water.

There is also the matter of warmth. During the time that life has existed on earth, the sun has become 25 per cent hotter, yet the mean temperature at the earth's surface has remained always fairly constant. Unlike Venus, which simply went on heating up till it reached temperatures far above what makes life possible, the earth gradually consumed much of the blanket of green-house gas – mostly carbon dioxide – which had originally warmed it. Feedback from living organisms seems to have played a crucial part in this steadying process and to have ensured, too, that it did not go too far. In this way the atmosphere remained substantial enough to avoid the fate of Mars, whose water and gases largely streamed away very early, leaving it unprotected against the deadly cold of space. Here again, conditions on earth stabilised in a most remarkable way within the quite narrow range which made continued life possible.

Lastly, there is the soil. We think of the stuff we walk on as *earth*, the natural material of our planet, and so it is. But it was not there at the start. Mars and Venus and the Moon have nothing like it. On them there is only what is called *regolith*, naked broken stone and dust. By contrast our soil, as Lynn Margulis points out, is a museum of past life:

> Soil is not unalive. It is a mixture of broken rock, pollen, fungal filaments, ciliate cysts, bacterial spores, nematodes and other microscopic animals and their parts. *'Nature,'* Aristotle observed, *'proceeds little by little from things lifeless to animal life in such a way that it is impossible to determine the exact line of demarcation.'* Independence is a political, not a scientific term.
>
> (Lynn Margulis and Dorion Sagan, *What is Life?*, London, Weidenfeld 1995, p. 26; emphasis mine)

In short, if all this is right, living things – including ourselves – and the planet that has produced them form a continuous system and act as such. Life, then, has not been just a casual passenger of the earth's development. It has always been and remains a crucial agent in determining its course.

PUTTING LIFE TOGETHER

I cannot discuss the scientific details further here. Orthodox scientists, though they were at first sceptical about it, now accept this general approach as one which can be used and debated within science.[4] Their disputes about these aspects of it will of course go on. But, as I have suggested, the importance of the concept is by no means confined to science. It concerns the general framework of our thought. It supplies an approach which, once fully grasped, makes a profound difference, not just to how we see the earth but to how we understand life and ourselves. The new scientific arguments bring back into focus the traditional imaginative vision of a living earth which I mentioned at the start – a vision which is already returning but needs to be made much clearer – and show how much we need this vision in our social and personal thinking.

As Lewis Thomas has pointed out, this vision has already dawned on many of us when we first saw the pictures of earth sent back by the astronauts:

> Viewed from the distance of the moon, the astonishing thing about the earth, catching the breath, is that it is alive. The photographs show the dry, pounded surface of the moon in the foreground, dead as an old bone. Aloft, floating free beneath the moist, gleaming membrane of bright blue sky, is the rising earth, the only exuberant thing in this part of the cosmos. If you could look long enough, you would see the swirling of the great drift of white cloud, covering and uncovering the half-hidden masses of land. If you had been looking a very long, geologic time, you could have seen the continents themselves in motion, drifting apart on their crustal plates, held aloft by the fire beneath. It has the organised, self-contained look of a live creature, full of information, marvellously skilled in handling the sun.
>
> (Lewis Thomas, *The Lives of a Cell*, London, Futura, 1976, p. 170)

(No other planet, incidentally, has continental drift and it seems that life may well have played a part in making this possible.)

THE PREVALENCE OF INTELLECTUAL APARTHEID

We will consider later what is involved in the use of the term 'life' on this planetary scale. For the moment, it seems important to consider how such suggestions made in science affect the rest of our thought. The scientific details that now articulate this picture of the living earth give it a new kind of standing because of the special importance that science has for us today. They make us bring our official scientific beliefs together with our imaginative life. As we have seen, that rapprochement is surely welcome but it is not easy for us. Many dualisms in recent thought have urged us to keep these matters apart. We are used to hearing of a stark war between the two cultures and of a total separation between facts and values. In our universities, the arts block and the science block tend to be well separated. But we will never make much sense of life if we do not somehow keep our various faculties on speaking terms with one another.

Much of the difficulty about grasping the concept of Gaia is not scientific but comes from this fragmented general framework of our thought. It arises – for scientists as well as for the rest of us – from these artificial fences that we have raised across the scene and centrally from Descartes' original fence between mind and body. Our moral, psychological and political ideas have all been armed against holism. They are both too specialised and too atomistic. As many people are pointing out today, that slant is giving us trouble in plenty of other places as well as over Gaia. Yet we find it very hard to change it.

ONE AQUARIUM, MANY WINDOWS

This difficulty in changing concepts is, of course, a common one. We are always in trouble when we are asked to think about the world in a new way. I have suggested that it is as if we had been looking into a vast, rather ill-lit aquarium through a single window and are suddenly told that things look different from the other side.

We cannot have a single comprehensive view of the whole aquarium – a single, all-purpose, philosophic theory of everything. Many prophets, from the seventeenth century to the nineteenth, from Leibniz to Hegel and Marx, have tried to give us such a view. But their efforts have proved misguided. The world is simply too rich for such reductive strait-jacketing. There is not – as Leibniz hoped – a single underlying quasi-mathematical language into which the views from all aspects can be translated.

This does not mean that no understanding is possible. We can relate these various aspects rationally because they all occur within the framework of our lives. We can walk round and look at other windows and can discuss them with each other. But we cannot eliminate any of them. We have to combine a number of different ways of thinking – the views through several windows, historical, biological, mathematical, everyday and the rest – and somehow fit them together.

When Galileo first expressed his views about the world, not only the Pope but the scientists of his day found them largely incomprehensible. Yet those ideas, when developed by Descartes, Newton, Laplace and the rest, shaped the set of windows through which the whole Enlightenment looked into the vast aquarium which is our world. That is the set through which many in our own age still want to see everything. This set is now called 'modern' by those who want to use that word more or less as a term of abuse for past errors, contrasting it with various 'post-modern' sets which may be expected to replace it. Though I don't myself find this vague time-snobbery very helpful, there is no doubt that the Cartesian vision has become quite insufficient.

THE AGE OF ALIENATION

As many people have pointed out,[5] the central trouble is the dualism of mind and body. The notion of our selves – our minds – as detached observers or colonists, separate from the physical world and therefore from each other, watching and exploiting a lifeless mechanism, has been with us since the dawn of modern science (and of the industrial revolution). Descartes taught us to think of matter essentially as our resource – a jumble of material blindly interacting. Animals and plants were machines and were provided for us to build into more machines.

It is this vision that still makes it so hard for us to take seriously the disasters that now infest our environment. Such a lifeless jumble would be no more capable of being injured than an avalanche would. Indeed, until quite lately our sages have repeatedly urged us to carry on a 'war against Nature'.[6] We did not expect the earth to be vulnerable, capable of health or sickness, wholeness or injury. But it turns out that we were wrong: the earth is now unmistakably sick. The living processes (or, as we say, 'mechanisms') that have so far kept the system working are disturbed, as is shown, for instance, by the surge of extinctions.

Descartes' world-view did, of course, produce many triumphs. But it produced them largely by dividing things – mind from body, reason from feeling, and the human race from the rest of the physical universe. It produced a huge harvest of local knowledge about many of the provinces. But it has made it very hard for people even to contemplate putting the parts together.

For a long time now our culture has tolerated this deprivation. But it has become a serious nuisance in many areas of knowledge. The rise of systems theory and complexity theory are thriving attempts to break its restraints. Another such place is the lively debate now going on about problems of consciousness – a topic once systematically tabooed by academics, but now agreed to constitute one of their most potentially interesting areas of study.[7] This change has been an intriguing showcase for the workings of intellectual fashion and it has interesting implications for discussions of Gaia. It is clear by now that many of us want to see our aquarium – our world, including ourselves – more as a whole, indeed, that we desperately need to do this. To do so, we must attend to aspects of it which Enlightenment dualism cannot reach, aspects which simply do not appear at our traditional window.

18

GODS AND GODDESSES

The role of wonder

WHY 'GAIA'?

One of these areas that has been made artificially difficult – the connection between scientific thought and the rest of life – comes out quaintly in the sharp debate about the implications of the name Gaia itself. That name arose when Lovelock told his friend, the novelist William Golding, that people found it hard to grasp his idea, and Golding promptly replied 'Why don't you call it Gaia?', which is the name of the Greek earth-goddess, mother of gods and men. That name, when he used it, did indeed rouse much more interest in the theory. Many people who had not previously understood it now grasped it and thought it useful. Others, however, particularly in the scientific establishment, now rejected it so violently that they refused to attend to the details of it altogether.

In our culture at present, people find it somewhat surprising that an idea can be large enough to have both a scientific and a religious aspect. This is because, during the last century, our ideas of religion, of science and indeed of life have all become narrowed in a way that makes it difficult to get these topics into the same perspective. (Here our window has become a good deal narrower than it was when Galileo and Newton and Faraday used it. They never doubted that these things belonged together.[1]) To get round this difficulty, Lovelock used a different image. He launched *the medical model of Gaia* – the idea of the damaged earth as a patient for whom we humans are the only available doctor, even though (as he points out) we lack the long experience of other sick planets which a doctor attending such a case really ought to have. So he invented the name *geophysiology* to cover the skills needed by such a physician.[2]

This medical imagery at once made it much easier for scientists to accept the notion of Gaia. When the point is put in medical terms, they begin to find it plausible that the earth does indeed in some way function as an organic whole, that its climate and oceans work together with living things to maintain a normal balance, and that what gravely upsets any part of the system is liable to upset others. They can see that, for such a whole, the notion of *health* is really quite suitable. And of course they find the patient Gaia, lying in bed and politely awaiting their attention, much less threatening than that scandalous pagan goddess.

SCIENTIFIC STATUS AND THE ISSUE OF GENDER

Lovelock, accordingly, came under great pressure to calm the scientists by withdrawing the goddess and for a while he seriously considered doing so. Eventually, however, he decided that the whole idea had to be kept together because the complexity was real. As Fred Pearce put it in an impressive article in *New Scientist*:

> Gaia as metaphor: Gaia as a catalyst for scientific enquiry: Gaia as literal truth: Gaia as Earth Goddess. Whoever she is, let's keep her. If science cannot find room for the grand vision, if Gaia dare not speak her name in *Nature*, then shame on science. To recant now would be a terrible thing, Jim. Don't do it.
>
> (Fred Pearce, 'Gaia, Gaia, don't go away' in *New Scientist*, 28 May 1994)

Lovelock didn't. He does indeed constantly emphasise the scientific status of the concept:

> I am not thinking in an animistic way of a planet with sentience ... I often describe the planetary ecosystem, Gaia, as alive because it behaves like a living organism to the extent that temperature and chemical composition are actively kept constant in the face of perturbations ... I am well aware that the term itself is metaphorical and that the earth is not alive in the same way as you or me or even a bacterium.[3]

But he still writes, with equal firmness: 'For me, Gaia is a religious as well as a scientific concept, and in both spheres it is manageable. . . . God and Gaia, theology and science, even physics and biology are not separate but a single way of thought.'[4]

This raises the question: is religious talk actually incompatible with science? It is interesting to note that in one prestigious area of science – an area which is often viewed as the archetype of all science – such talk is readily accepted. That area is theoretical physics. As Margaret Wertheim has pointed out, most of the great physicists of the past, from Copernicus to Clerk Maxwell, insisted that their work was primarily and essentially religious. Rather more remarkably, their modern successors still make the same claim. As she says:

> In spite of the officially secular climate of modern science,

physicists have continued to retain a quasi-religious attitude to
their work. They have continued to comport themselves as a
scientific priesthood, and to present themselves to the public
in that light. To quote Einstein, 'A contemporary has said, not
unjustly, that *in this materialistic age of ours the serious scientific
workers are the only truly religious people*'.

(Margaret Wertheim, *Pythagoras' Trousers*, London,
Fourth Estate, 1997, p. 12; emphasis mine)

Einstein himself showed how seriously he took this thought by
constantly referring to God in explaining his own reasoning
('God does not play dice', . . . 'The Lord is subtle but not mali-
cious' and so forth). And he explicitly said that he meant it:
'Science can only be created by those who are thoroughly
imbued with the aspiration towards truth and understanding.
The source of this feeling, however, springs from the sphere of
religion (Einstein, 'Science and Religion' in *Nature*, vol. 146,
no. 65, 1940, p. 605).

Later physicists might have been expected to dismiss this
approach as a mere personal quirk of Einstein's, but they have
not. Instead, many of them have developed it in best-selling
books with titles such as *God And The New Physics*,[5] *The Mind Of God*,[6]
The God Particle,[7] *The Physics of Immortality: Modern Cosmology, God and the
Resurrection of the Dead*[8] and many more.

Is there perhaps some special reason why religious talk of this
kind can count as a proper language for physics, but becomes
inappropriate and scandalous when the chemical and biological
concerns of Gaian thinking are in question? Or is it perhaps not
so much the subject-matter as the sex of the deity that makes the
scandal? Is it perhaps held to be scientifically proper to speak of a
male power in the cosmos but not of a female one? There is a
powerful tradition which might make this odd view look plausible. As Wertheim shows, throughout the history of physics, a
strong and somewhat fantastic element of misogyny has indeed

accompanied the sense of sacredness that always distinguished this study. The physical priesthood was a male one guarding a male god and it went to great lengths to protect its secrets from intruding females:

> Walter Charleton, another founding member of the Royal Society, summed up many of his colleagues' antipathy towards women when he wrote, 'you are the true Hienas that allure us with the fairness of your skins ... You are the traitors to wisdom, the impediments to industry ... the clogs to virtue and the goads that drive us all to Vice, Impiety and Ruin'. Henry Oldenburg, the Society's first secretary, declared that its express purpose was 'to raise a Masculine philosophy'.... This bastion of British science did not admit a woman as a full member until 1945.[9]

This talk of 'a Masculine philosophy' echoes, of course, Francis Bacon's clarion-call for the new science to produce 'a Masculine birth of time' where men could turn their 'united forces against the nature of things, to storm and occupy her castle and strongholds'.[10]

THE VALUE OF WONDER

Of course the personifications in thinking of this kind should not be taken literally. Yet the reverent, awe-struck attitude that lies behind those personifications is surely a suitable one both for science and for our general relation to the cosmos. Einstein was not being silly. Anyone who tries to contemplate these vast questions without any sense of reverence for their vastness simply shows ignorance of what they entail. And of course, if the system of life itself is taken to have participated in the history of evolution in the sort of way that Gaian thinking suggests, then a substantial part of this reverence is surely due to that system. If it has indeed played a crucial part in stablising conditions on earth through billions of years, to the point where we ourselves are now here and able to profit from them – if it has managed the remarkable feat of preserving the atmosphere and controlling the temperature, thus saving the earth from becoming a dead planet like Mars and Venus and turning it instead into the cherished blue-green sphere whose picture we all welcomed – if it has done all this for us, then the only possible response to that feat is surely wonder, awe and gratitude.

That sense of wonder and gratitude is clearly what the Greeks had in mind when they named the earth Gaia, the divine mother of gods and men. They never developed that naming into a full humanisation. They never brought Gaia into the scandalous human stories that they told about other gods – stories which, in the end, made it impossible to take those gods seriously at all. But the name still expressed their awe and gratitude at being part of that great whole.

And today there is evidently more, not less, reason to feel that awe and gratitude, because we have learnt something of the scope of the achievement. The sense of life itself as active and effective throughout this vast development has become stronger, not weaker, with our understanding of our evolutionary history. This is the sense that Darwin expressed when he wrote, at the end of the The Origin of Species, 'There is grandeur in this view of life'.

INTRINSIC VALUE AND THE SOCIAL CONTRACT

It does not seem to me to matter much whether one calls this wonder and reverence *religious* or not except to people who have declared a tribal war about the use of that word. It is of course an element that lies at the root of all religions. In the great religions with which we are familiar, it always plays its part and is subsumed within a wider whole. Reverence for the creation can there quite properly inspire and enrich the reverence that is due to its creator.

But such wonder and reverence are equally essential to belief-systems that reject religion. All such systems involve some order of values, some pyramid of priorities which has to end somewhere. In order to make sense of our lives, we have to see some things as mattering in themselves, not merely as a means to something else. Some things have to have what the theorists call intrinsic value. Secular thought in the West has not dropped that notion. Instead, during the last century, it has simply decreed that human individuality itself is the only thing that has this status. Today it uses words such as *sacred* and *sanctity* readily to describe human life, but becomes embarrassed if they are used for anything else. People with this approach tend to be alarmed by the direct reverence for the non-human world that was expressed by people like Wordsworth and Rousseau and to treat it as something not quite serious.

Here we come back to the question that I mentioned at the outset about the possible reasons why the fate of the earth should concern us. The early twentieth century's humanistic creed that only people have value – that non-human affairs do not matter except for their effect on people – means that there cannot be any such reason. This is the unspoken creed that leaves us – or at least leaves the professional moralists among us – so puzzled by the environmental crisis – by the thought that we might actually owe some direct duty to the biosphere.[11]

Our individualism has accustomed us to using a minimalist moral approach which gives us no clue to such matters. But that minimal approach has, of course, already got a difficulty in explaining why each of us should be concerned about any other individual besides our own self in the first place – why our value-system should ever go beyond simple egoism. As we have seen, it answers this question in terms of the social contract which is supposed to make it worth while for each of us to secure the interests of fellow citizens. The answer to the question 'Why should I bother about this?' is then always 'Because of the contract which gives you your entrance-ticket to society'.

This contract model excludes dealings with anything non-human. It works quite well for political life within a nation – for which, of course, it was originally invented. But even there it leaves out most of life. Even within our own lives, we know that we cannot think of rights and duties as optional contracts set up between essentially separate individuals. Relations between parents and children are not like this, and each of us, after all, started life as a non-contracting baby. Nor indeed are most of our personal relations. But we have not yet grasped how much worse this misfit becomes when we have to deal with the rest of the world.

Even over animals, the legalistic notion of contractual rights works badly. And when we come to such chronic non-litigants as the rain forest and the Antarctic it fails us completely. If duties are essentially contractual, how can we possibly have duties to such entities? John Rawls raised this question rather suddenly as an afterthought at the very end of his famous book *A Theory of Justice* and could only say that it was one which lay outside his contractual theory.[12] He added that it ought to be investigated some day. But, as often in such cases, the real response has to be 'you shouldn't have started from here'. Rawls' book was the definitive statement of contract ethics and it marked the end of the era when they could pass as adequate.

GRANTING CITIZENSHIP TO WILDERNESSES

Individualism is bankrupt of suggestions for dealing with these non-human entities. Yet we now have to deal with them, and promptly. They can no longer be ignored. Clearly, too, most of us do now think of the human drama as taking place within this larger theatre, not on a private stage of its own. The Darwinian perspective on evolution places us firmly in a wider kinship than Descartes or Hobbes ever dreamed of. We know that we belong on this earth. We are not machines or alien beings or dis-embodied spirits but primates – animals as naturally and incur-ably dependent on the earthly biosphere as each one of us is dependent on human society. We know that we are members of it and that our technology already commits us to acting in it. By our pollution and our forest-clearances we are already doing so.

What element, then, does the concept of Gaia add to this dawning awareness? It is something beyond the fact of human sociability, which has already been stated, for instance by com-munitarians. It is not just the mutual dependence of organisms around us, which is already to some extent being brought home to us by ecology. It goes beyond thinking of these organisms as originally separate units that have somehow decided or been forced to co-operate – as basically independent entities which drive bargains for social contracts with each other ('reciprocal altruism') because they just happen to need each other in order to survive. The metaphysical idea that only individuals are real entities is still present in this picture and it is always misleading. *Wholes and parts are equally real.*

Recent habits make it hard for us to take this in. As a lot of science fiction makes clear, we are still amazingly ready to think of our species as a mere chance visitor on this planet, as some-thing too grand to have developed here. Of course it is true that we are a somewhat special kind of primate, one that is particu-larly adaptable through culture and gifted with singular talents. But those gifts and talents still come to us *from the earth* out of which we grow and to which we shall return. The top of our tree still grows from that root as much as the lower branches. We cannot live elsewhere. Our fantasies of moving to outer space mean no more than the magic tales with which other cultures have often consoled themselves for their mortality. Even people who still expect that move in the long term are beginning to see that it cannot be expected to arrive in time to relieve our present emergency. Since the end of the Cold War, NASA finds it increasingly hard to raise funds to keep space programmes going. And environmental disasters are likely to make that process harder, not easier.

All this means that, in spite of recent influences, direct concern about destruction of the natural world is still a natural, spontaneous feeling in us and one that we no longer have any good reason to suppress. Most people, hearing about the wanton destruction of forests and oceans find it shocking and – as has become clear in the last few decades – many of them are prepared to take a good deal of trouble to prevent it. This feeling of shock and outrage is the energy-source which makes change possible.

It has not, of course, been properly tapped yet. As happened over nuclear power, it takes a disaster to bring such needs home to people. Yet the feeling is there and it is surely already becoming stronger and more vocal. It is, of course, what leads people to subscribe to organisations trying to protect the environment. Though we have been educated to detach ourselves from the physical matter of our planet as something alien to us, this detachment is still not a natural or necessary attitude to us. Since we now know that we have evolved from a whole continuum of other life-forms and are closely akin to them – a point which nobody ever explained to Descartes – it is not at all clear why we should want to separate ourselves from them in this way. On this point, of course, the findings of modern science agree much better with the attitude of those supposedly more primitive cultures where people see themselves as part of the whole spectrum of life around them than they do with the exclusive humanism of the Enlightenment. They also agree better with most of our everyday thought. The element in that thought which is now beginning to look arbitrary and unreal is its exclusive humanism.

19

WHY THERE IS SUCH A THING AS SOCIETY

THE SURPRISING INEFFICIENCY OF SELFISHNESS

Indignant concern on behalf of the environment does, then, already exist. Our difficulty is that we cannot see how to fit it into our traditional morality which – both in its Christian and its secular forms – has in general been carefully tailored to fit only the human scene.

How should we deal with this conceptual emergency? I do not think that it is very helpful to proceed as some moralists have done by promoting various selected outside entities such as 'wildernesses' to the status of honorary members of human society. If we claim (for instance) that a wilderness such as the Antarctic has intrinsic value because it has independent moral status, meaning by this that we have decided to grant it the privilege of treating it like an extra fellow citizen, we shall sound rather inadequate. These larger wholes are independent of us in a quite different sense from that in which extra humans – or even animals – who were candidates for citizenship might be so. Our relation to them is of a totally different kind from the one which links us to our fellow citizens.

There is, indeed, something unreal about the whole way of thinking which speaks of these places as though they were distinct individual 'wildernesses', units which are applying separately for admission to our value-spectrum. Though we divide them for our thought, they function as parts of the whole. At present, indeed, the Arctic and the Antarctic are letting us know this because their ice, melted by global warming, is affecting the entire state of the oceans. That process is already producing floods which threaten the destruction of places such as Bangladesh and Mauritius and widespread damage elsewhere. Nearer home, it also looks liable to upset the Gulf Stream in a way that may drastically chill the climate of Europe. Without that convenient warming system, we in Britain would find ourselves ten degrees colder, sharing the climate of Labrador, which is on much the same latitude. And if that change happens it could apparently happen quite quickly. Globalisation is no longer a distant option. It is here already.

This is, of course, a prudential consideration. It may suggest that rational self-interest alone will be enough to guide us here – as Hobbes supposed it always would be. And of course it is true that self-interest should indeed drive us this way. The odd thing is that it does not.[1] The human imagination does not work that way. When things go well, we simply don't believe in disasters. Long-term prudence, reaching beyond the accepted, routine precautions of everyday life, is therefore an extraordinarily feeble motive.

Prudence is supposed to deal in probabilities as well as in certainties. And the increasing probability of environmental disaster has been well-attested for at least the last thirty years. During all that while, every time that the travellers in steerage pointed out that the ship was sinking the first-class passengers have continued to reply placidly, 'Not at our end'. Only very gradually and shakily is this prospect beginning to be admitted as an influence on policy – a topic that should be allowed now and then to compete for the attention of decision-makers, alongside football and teenage sex and the Dow-Jones Index and European Monetary Union. Only gradually is it beginning to emerge that ecology is actually a more important science than economics – that the profitable exchange of goods within the ship is a less urgent matter than how to keep the whole ship above water. When the story of our age comes to be written, this perspective may surely seem surprising.

Our imaginations, however, are not necessarily ruled by our reason. We do not easily expect the unfamiliar, and major disasters are always unfamiliar. When we are trying to be prudent, our thoughts turn to well-known and immediate dangers, nervously avoiding a wider scene. That is why self-interest alone cannot be trusted to answer our question about why the earth should concern us. Of course prudence must come in, but unless other reasons are already recognised prudence usually manages to evade the larger topic. That is why we need to think about those other reasons – about the ways in which the terrestrial whole, of which we are a part, directly concerns us, and would still do so even if we could get away with abusing it. As I am suggesting, we shall never grasp the nature of that kind of concern so long as we try to model it on the civic concern that links fellow citizens. *Duties to wholes, of which one is a part, naturally differ in form from duties to other individuals.*

OUTWARD- AND INWARD-LOOKING CONCERNS

Since the Enlightenment, our culture has made huge efforts to exclude outward-looking duties from Western morality. Pronouncements such as 'there is no such thing as society' and 'the state is only a logical construction out of its members' are only recent shots in this long individualist campaign. But the natural strength of outward-looking concern can be seen from the way in which many such duties are still accepted. For instance, the idea of *duty to one's country* still persists and it certainly does not just mean duty to obey the government. The ideas of *duty to a family, clan, locality or racial group* still have great force, even in our society where they have been deliberately played down, whenever one of these groups feels threatened by outside oppression. The current revival of nationalism among various groups all over the world, and the emphasis laid on *sisterhood* by feminists, all testify to this force. In other cultures, where no attempt has been made to undermine it, its strength is unmistakable.

Another corporate claim which can operate powerfully is the idea of a *duty to posterity*. This is not just the idea of a string of separate duties to particular future individuals. It is rather the sense of being part of a great historical stream of effort within which we live and to which we owe loyalty. That identification with the stream explains the sense in which we can – rather surprisingly – owe duties to the dead and also to a great range of anonymous future people, two things which have baffled individualistic thinkers. Even when there is no conscious talk of duty, people who work in any co-operative enterprise – school, firm, shop, orchestra, theatrical company, teenage gang, political party, football team – find it thoroughly natural to act as if they had a duty to that enclosing whole if it is in some way threatened.

And this, it seems to me, is what is now beginning to happen about the earth itself, as the threat to it begins to be grasped. When an enclosing whole which has been taken for granted is suddenly seen as really endangered, all at once its hidden claims become visible. It would be good if we could accept the overwhelming existing evidence of a terrestrial emergency without needing to be hit by a direct disaster. But whatever causes that belief to be accepted, once it becomes so there is surely little doubt about the duty it lays upon us.

STATES AND ORGANISMS

It is not surprising that our mainstream tradition has played down this corporate element in morals. Political theorists such as Hobbes, Locke and Rousseau – and their contemporaries in active politics – wanted above all to stop certain dominant groups, notably in the churches, from exploiting this loyalty for their own ends. They succeeded to an extent which would surely have astonished them if they could have foreseen it, and which Rousseau at least would have found alarming. Between them, they managed to swing the balance of moral thinking right over to its individualistic pole.

As we have seen, they did not manage to destroy the idea of corporate duty entirely. *Fraternity* was supposed to be among the ideals of the French Revolution, though in practice it was usually thrust aside by equality and freedom. Rousseau himself did try to balance the individualism of his contract theory by introducing the idea of the general will, a corporate will in the nation distinct from the mere summing of separate decisions – something to be relied on more deeply, something which individuals should seek out and follow. This and similar hints were developed by Hegel into a fully fledged organic theory of the state, by which individuals are always incomplete entities, more or less comparable with cells in a plant or animal, needing to find their place in a wider whole for full self-realisation.

Up to a point this suggestion clearly has to be true. Most of us, if we can act freely at all, want to place ourselves within such larger groupings – families, clubs, friendships, orchestras, gangs, political movements. But it is a sort of doctrine which sounds very different according to which kind of larger group we have in mind. By bad luck, Hegel centred his theory on the nation state and in particular on his own state of Prussia, which was then (in the early nineteenth century) preparing to dominate the rest of Germany and thereby the rest of Europe. Marx, following Hegel's organic approach, also expected his precepts to be taken up in Germany and, though he envisaged a distant time when nation states would not be needed, he expected them to be the main social unit for the foreseeable future. As the eventual adoption of Marxism in Russia did not produce any sort of utopia, it is not surprising that these two unattractive examples have put people off organic theories of society, or that many of them end up saying, with Nietzsche's Zarathustra: 'The State lieth in all the languages of good and evil: whatsoever it saith, it lieth: whatsoever it hath, it hath stolen' (Friedrich Nietzsche, *Thus Spake Zarathustra*, part 1, section 'Of The New Idol').

Thus, through most of the twentieth century, many prophets in the West preached a kind of narrow and romantic individualism, a moral outlook which simply assumes that individual freedom is the only unquestionable value. This is a doctrine held in common by J.-P. Sartre and Ayn Rand. Despite the difference of style, the European and the American forms of it share a central message – social atomism. Both conceive the individual's freedom as negative – a matter of avoiding interference. Politically, however, there is rather an important difference because the kind of entity that counts as 'an individual' is different in the two versions.

The European version still speaks of individual people and therefore stays close to real anarchism. The American one, however, expands to include 'commercial freedom'. And commercial freedom, in its modern form, is a different thing and a very strange one. The entities which it conceives as free are no longer individuals but corporations, often very big and impersonal ones. The rhetoric of 'free trade', in fact, does not now refer to individual freedom at all. The old romantic vision of commercial freedom which (as we shall see in a moment) Herbert Spencer presented in the 1880s – a vision of heroic individual tycoons carving out the course of evolution with their bare hands – does not fit today's conditions at all, whatever may be thought of its exactness in his own day.

There has, in fact, been an extraordinary shift here in the central tenet of individualism. The metaphysical belief in human individuals as the true atoms of social life – the only properly real and sacred kind of unit – has given way. At the moment, the focus has shifted to another kind of entity, the big corporation. But since that kind of entity, in its turn, is now beginning to look rather less than ultimate – since the Internet is threatening its supremacy by building a more diffused way of doing business, while individual speculators infest it from within and shake its control – this does not seem likely to be the end of the story. These corporations may prove to be dinosaurs, entities remembered only as we remember mediaeval guilds. What surely emerges is that the whole idea of a single favoured, exclusively real unit was mistaken in the first place. *Life goes on on various scales, each of which is real and has to be thought of in its own terms.*

SOCIALITY SURVIVES

This shift of emphasis to a kind of corporate freedom is, however, just one more indication of how – as communitarians have recently been pointing out – individualist propaganda cannot destroy the corporate element in morals. Of course we still value our personal freedom very highly. Psychologically, perhaps we value it the more because of the stress produced by overcrowding, by the sheer increase in human numbers and in social mobility during the past century. We all see far more people, especially far more strangers, in our daily lives than our ancestors did, which certainly imposes stress and social exhaustion.

Yet humans – even modern, civilised humans – are still social animals to whom, on average, the desolation of loneliness is a much worse threat than the interference of their fellows. On the positive side, too, we have talents and capacities which absolutely require generous, outgoing co-operation for their fulfilment – a point which Hegel got right. Paradoxically, there are many things which a free, solitary individualist is not free to do. He cannot be a parent, a quartet-player, a tragic actor, a teacher, a social reformer or even a revolutionary. Even Nietzsche's Zarathustra noticed this difficulty: 'A light hath dawned on me. I need companions . . . living companions which follow me because they desire to follow themselves – and to go to that place whither I wish to go' (*Thus Spake Zarathustra*, Introductory Discourse, section 9).

In fact (as Bishop Butler pointed out against Hobbes) apart from certain narrow political contexts human beings are not in the least like the pure, consistent, prudent egoists that social contract thinking requires. And today people are coming to see this.

Of course it is true that we need to stop the powerful oppressing the weak, so we must have political institutions to prevent the exploitation of these corporate loyalties. That is why we need a free press to answer the propaganda of governments. And since the press itself comes under commercial pressure, that pressure, working through the labour market, through advertisements and through countless other channels, is, on the whole, much more alarming today than the power of religion. But the need to ward off these dangers cannot mean that we can do without corporate loyalties altogether. The outgoing, social side of human life vitally needs them.

20

PARADOXES OF SOCIOBIOLOGY AND SOCIAL DARWINISM

THE EXALTATION OF COMPETITION

I suggested earlier that one thing which makes this hard to see today is the intensely individualistic ideology which pervades recent sociobiological discussions of evolution – their suggestion that organisms evolve so exclusively by competition that co-operation at any level is not just wrong but impossible, since it is contrary to nature.

That story is officially based on Darwin's work, but is actually much more extreme than anything to be found there. Its real ancestor is Herbert Spencer. The remarkable thing about it is the unbalanced rhetoric in which it is expressed, the lurid imagery used to inflate an interesting but modest range of facts about natural selection into an all-purpose individualistic melodrama.

This rhetoric is distinct from the official purpose of these sociobiological discussions and often conflicts with it. Officially they are an entirely admirable celebration of evolution and of our oneness with the rest of nature. Sociobiological thinking aims to counter the narrow, exclusive humanism of which I have been complaining by making humans appear on their proper scale in the vast evolutionary context. It aims to show the grandeur of that context in a way that removes any sense of degradation from our assimilation to the rest of nature and makes us feel at home in the natural world.

I entirely accept all this, so I shall say no more of it here. But the language used to express it, and the implications that follow that language, have a very different effect – an effect that bears sharply on our central topic of individualism. It is worth following it out further. The spiritual ambitions of this sociobiological thinking are high. As Edward O. Wilson puts it:

> I consider the scientific ethos superior to religion . . . The core of scientific materialism is the evolutionary epic . . . the evolutionary epic is probably the best myth we will ever have. It can be adjusted until it is as close to truth as the human mind is constructed to judge the truth. And if that is the case, the mythopoeic requirements of the mind must somehow be met by scientific materialism so as to reinvest our superb energies . . . Every epic needs a hero, the mind will do . . . Scientific materialism is the only mythology that can manufacture great goals from the sustained pursuit of pure knowledge.
>
> (E.O. Wilson, *On Human Nature*, Cambridge, Harvard University Press, 1978, pp. 201, 203, 207)

This is a lofty hope. It seems strange, then, that most sociobiological accounts of the workings of nature sound rather like descriptions of a bad day on the New York Stock Exchange. Thus M.T. Ghiselin, echoing Dawkins' and Wilson's doctrines in language which is only slightly more lurid than theirs, writes that:

> The evolution of society fits the Darwinian paradigm in its most individualistic form. The economy of nature is competitive from beginning to end ... the underlying reasons for social phenomena are manifest. They are the means by which one organism gains some advantage to the detriment of another.
>
> No hint of genuine charity ameliorates our vision of society, once sentimentalism has been laid aside. What passes for cooperation turns out to be a mixture of opportunism and exploitation. The impulses that lead one animal to sacrifice himself for another turn out to have their ultimate rationale in gaining advantage over a third: and acts 'for the good' of one society turn out to have been performed for the detriment of the rest. Where it is in his own interest, every organism may reasonably be expected to aid his fellows. Where he has no alternative, he submits to the yoke of communal servitude. Yet, given a full chance to act in his own interest, nothing but expediency will restrain him from brutalising, from maiming, from murdering – his brother, his mate, his parent or his child. Scratch an 'altruist:' and watch a hypocrite bleed.
>
> (M.T. Ghiselin, *The Economy of Nature and the Evolution of Sex*, Berkeley, Cal., University of California Press, 1974, pp. 155–6)

This strange conclusion is reached by three steps – all of them part of the general sociobiological approach. First, the innumerable altruistic and co-operative activities which are well known to occur among plants and animals are treated as if they were all just devious stratagems to produce more descendants. Second, it is implied that producing these descendants is itself somehow an advantage to the parent, an advantage described as an increase in its 'inclusive fitness', which simply means 'having more descendants'. Third and strangest of all, lurid motive-words such as 'selfishness', 'spite' and 'charity' are used to dramatise that ancestor's situation by making it seem to be deliberately planning its dynastic future. Things are then further confused by sometimes – but not always – attributing these motives to the genes rather than to the organisms themselves.

By accepting all these odd moves, sociobiologists are able to conclude that the blackbird which gets eaten by the cat in trying to protect its young has actually pulled off a selfish coup, because (since its young survive) it has arranged to have more descendants than the blackbird next door, which didn't do so. Alternatively, if the selfishness is attributed to the gene, the bird itself is merely a robot, a helpless puppet used to maximise the spread of its genes. Both these interpretations enable the authors to feel confident that they have shown that ordinary affection (something which is clearly an unbearable embarrassment to them) cannot be a real determinant of behaviour.

What is the word 'selfish' doing here? Officially, in these writings, it is supposed to be a harmless technical term referring to genes and meaning something like 'selectable, able to be selected alone'. The authors, however, constantly oscillate between this technical meaning and the ordinary sense of the word – a sense which, with a word so common and emotive, cannot really be thrown off. Thus Richard Dawkins, having started his book by saying that 'we are survival machines – robot vehicles blindly programmed to preserve the selfish molecules known as genes . . . We and all other animals are machines created by our genes' quickly begins to describe this 'selfishness' of the genes as if it were a motive by qualifying it with words like 'ruthless', and shortly goes on to warn us that this motive – selfishness in the ordinary sense – actually belongs to ourselves: 'if you wish, as I do, to build a society in which individuals co-operate generously and unselfishly towards a common good, you can expect little help from biological nature. Let us try to teach generosity and altruism, because *we are born selfish*' (Richard Dawkins, *The Selfish Gene*, Oxford, Oxford University Press, 1976, pp. x, 2, 3 and 215; emphasis mine) – advice which he repeats solemnly in the last paragraph of his book.

These writers do, of course, occasionally explain that their words must not be taken literally. But the disclaimers are brief and so completely without effect on their surroundings that they have no more force than the tiny warnings on cigarette packets. Like those warnings, they are cancelled by their context. As in the Mafia, the sin is confessed but there is no intention of amendment. These writings continue to mix metaphor and literal science so inextricably that it is clear the authors themselves do not know how to distinguish them and it is not to be expected that their readers should do so.

SPENCERISM AND THATCHERISM

What accounts for this choice of words? Why did anybody ever pick on so odd a word as *selfish* as a technical term when something like *selectable* would have been so much more appropriate? And why did the public respond so eagerly, making these books instant best-sellers?

It is surely clear that the reason for all this had nothing to do with science. It lay in a fresh outcropping of the strong egoistic, individualistic strain in our political and moral thinking which dates from Hobbes and Locke. That strain became associated with evolution in Darwin's day through Herbert Spencer and the social Darwinists in a romantic glorification of capitalist enterprise entirely typical of that epoch.

Already at that time, the imagery had begun to give trouble. Theorists, including Darwin, who discussed conflicts of interest in the rest of nature constantly used images drawn from two particular human institutions – war and commerce. In that non-human context it was not too hard to remember that these were only metaphors. But when the discussion turned back to human affairs, it became much harder to be clear that these were not literal descriptions of human life, new truths about its characteristic motives and intentions, truths which showed that it could all be reduced to these two simple models. The drama shaped by those models was then projected back, in its turn, onto the whole cosmos, producing a picture of the universe in which commercial rivalry provided the key guiding principle for everything 'from gas to genius'.

Spencer and his followers thus saw competition as the all-explaining pattern both for human life and (somewhat casually, for they were not scientists) for the rest of nature. Darwin himself, always anxiously aware of the limits of our knowledge, carefully avoided these extensions. He just wanted to do biology. Though he naturally shared many of the social beliefs of his day, he had no faith in attempts to justify them by reference to the vast sweep of evolution.[1] Spencer, by contrast, was sure that simple rules united these topics. What evolution now demanded was (he explained) plenty of commercial freedom for capital-istic enterprises. In the 1880s he toured the United States, preaching the gospel that 'the millionaires are a product of nat-ural selection . . . It is because they are thus selected that wealth – both their own and that entrusted to them – aggregates under their hands'. His works accordingly outsold those of any other philosopher there for more than a decade.[2]

Since that time, the confusions of crude social Darwinism have, of course, been repeatedly answered and by the mid-twentieth century it had fallen into disfavour. But, as often hap-pens, the imagery and the temper behind it remained, waiting for another theoretical vehicle. In the United States, in con-sequence of Spencer's influence and the actual development of capitalism, a kind of faith in social Darwinism is still often taken for granted. (Wilson, for one, evidently accepts it as his back-ground.) In Britain this is less true, but romantic individualism itself still has a more general appeal which makes it surface from time to time in most of us as a handy simplification.

Dawkins' and Wilson's[3] books both came out in the mid-1970s, a time when, on both sides of the Atlantic, the moral tide was on the turn from the relatively idealistic, co-operative temper generated after the Second World War towards a more relaxed mood of self-expression and self-indulgence. In Britain, the real advantages which the Welfare State had produced were becoming familiar. They were beginning to be taken for granted while the drawbacks which had gone along with them began to be sharply felt. Bureaucratic control and the 'culture of dependence' were seen as grave evils. The immediate remedy prescribed for them was a return to commercial freedom and to extreme individualism generally, which was seen for a time, with a good deal of unrealistic nostalgia, as a social panacea.

Appearing at this point, these two exciting sociobiological bibles were seen as simple celebrations of selfishness and suited that temper perfectly. Their doctrines have therefore, not unnaturally, been described as biological Thatcherism. It is right to stress that the authors themselves reject this charge. When taxed with moral or political implications they usually recoil in astonishment saying that they are only doing science. And it is quite true that they do not make the kind of specific political recommendations that the social Darwinists made. But one does not have to give detailed advice to have a moral and political influence. No doubt their vagueness about moral contexts is genuine. But this only shows, I think, how poorly current scientific education prepares scientists to understand the crucial role of science in the rest of life.

As I remarked earlier, the patterns of thought used in science, and especially its dominant images, come in the first place from the life around and are often returned to that life in significantly altered forms. And today, when we respect science so deeply, it is increasingly natural to turn to those thought-patterns for guidance on central matters in our lives. Indeed, Dawkins himself introduces his book by offering just this kind of guidance:

We no longer have to resort to superstition when faced with *the deep problems: Is there a meaning to life? What are we for? What is man?* . . . [then, quoting a bizarre remark from G.G. Simpson] 'all attempts to answer that [last] question before 1859 are worthless and . . . we will be better off if we ignore them completely'.

(*The Selfish Gene*, p. 1)

He clearly recognises here, in fact, that an idea can indeed have both scientific and moral importance, which is what I am saying is true in the case of Gaia. But if this is so, then it seems that scientists are responsible for thinking through the social consequences of what they are suggesting. On these vast subjects, it is not possible only to be doing science. Science on this scale involves morals and politics as well.

21

MYTHOLOGY, RHETORIC AND RELIGION

THE EVOLUTIONARY CHURCH

I have cited the sociobiologists at some length in order to point out how in their case science, morals and politics are indeed combined and are, as I think, confused by the writers' failure to appreciate the relation between them. Sociobiology, however much it celebrates the unity of life, also tends to generate, through its uncontrolled rhetoric, a mindless social atomism. Gaian thinking, in which the relation between the two aspects is far better understood, can, I believe, do a good deal to correct that bias.

It is also interesting that the sociobiological celebration of evolution has, equally with Gaia, a religious angle. As we have seen, Wilson for one welcomes this idea. He describes the evolutionary story, and the materialism which he thinks underlies it, as a rival mythology in direct competition with traditional religion, an improved substitute which can be relied on to supersede it:

Religion constitutes the greatest challenge to human socio-biology and its most exciting opportunity to progress as a truly original theoretical discipline. If the mind is to any extent guided by Kantian imperatives, they are more likely to be found in religious feeling than in rational thought . . . Make no mistake about the power of scientific materialism. It presents the human mind with an *alternative mythology* that until now has always, point for point in zones of conflict, defeated traditional religion . . . The time has come to ask: Does a way exist to *divert the power of religion* into the great new enterprises that lay bare the source of that power?

   (E.O. Wilson, *On Human Nature*, Cambridge, Mass., Harvard University Press, 1978, pp. 175 and 192–3; emphases mine)

This mythology is, he says, 'guided by the corrective devices of the scientific method, addressed with precise and *deliberately affective appeal* to the deepest needs of human nature, and kept strong by the *blind hopes that the journey on which we are now embarked will be farther and better than the one just completed* (p. 209, my emphases). In short, it is a faith, and one carefully framed, in accord with an understanding of public relations, to suit the tastes of its potential congregation. This consideration is evidently strong. For it is hard to see what scientific grounds Wilson could possibly offer for expecting the future to be better than the past. The soothing expectation of incessant progress on earth simply grew up gradually in our culture as people abandoned the idea of relying on a heaven which would make it unnecessary. It did, of course, figure in Lamarck's and Herbert Spencer's view of evolution. But scientists are supposed now to have abandoned that. *Progress* of this kind forms absolutely no part of Darwin's doctrine and current science says nothing to support it.

   Wilson is, of course, recognising an important truth here. He has seen that every thought-system has at its core a guiding myth – not a myth in the sense of a lie but of an imaginative vision, a picture which does indeed 'express its appeal to the deepest needs of our nature'. And he sees, what not all scientists do see, that this is as true of world-views that are accepted as scientific as it is of other world-views. True beliefs need their imagery quite as much as false ones. A steady stream of imagery has in fact played a crucial part in the rise of modern science. In the early days that imagery centred on comparing the physical world with the clockwork *machines* of the early industrial revolution, an analogy which, after proving immensely useful, is now running into trouble in many places, notably in particle physics. Darwin, for his part, notoriously relied greatly on the metaphor of *selection*, another comparison which has been very useful but has proved to have drawbacks.

There is nothing wrong with such images, but, as these examples show, no one of them can ever serve for all purposes. No picture should be allowed to become an imaginative monoculture. They all need to be corrected sometimes by other ways of thinking. The mythology that is offered today as a celebration of evolution by people like Wilson and Dawkins is one-sided because it is profoundly and arbitrarily individualistic. Its imagery of *selfishness, spite* and *grudging, investment, cheats, war games* and the rest unmistakably reflects the naive social atomism of the 1970s and 1980s.

No doubt this dramatic language has been useful in bringing out certain aspects of evolution. No doubt it can still be used to investigate them further. But it really is important that people who use it should recognise its mythical character – should see that it is just one optional vision among others, a slanted, incomplete picture belonging to a particular epoch, a story which always needs others to correct it, not a final universal truth. The mythical quality which is often held to be an objection against the concept of Gaia is certainly no less present in the Selfish Gene.

In the real world, as many biologists have pointed out,

co-operation and competition go together as two sides of the same coin and, of the two, when things get at all complicated, co-operation must usually come first because it makes other interactions possible. If we consider how much co-operation is needed to organise even a competitive institution such as the stock-exchange – or, indeed, even to organise a single school sports day – this should surely be obvious. As Brian Goodwin puts it:

> There is as much co-operation in biology as there is competition. Mutualism and symbiosis – organisms living together in states of mutual dependency – such as lichens that combine a fungus with an alga in happy harmony, or the bacteria in our guts, from which we benefit as well as they – are an equally universal feature of the biological realm. Why not argue that 'co-operation' is the great source of innovation in evolution, as in the enormous step, aeons ago, of producing a eukaryotic cell, one with a true nucleus, which came about by the co-operation of two or three prokaryotes, cells without nuclei?
>
> (Brian Goodwin, *How the Leopard Changed its Spots*,
> London, Orion Books, 1994, pp. 166–8)

As he points out, these co-operations are particularly striking at the microbial level, which is why recent increases in understanding of that level have directed people's attention to them. But at every level they are an essential feature of life. Modern Darwinism, he says, describes

> the evolutionary process as one driven by competition, survival and selfishness. This makes sense to us in terms of our experience of our own culture and its values ... Darwinian metaphors are grounded in the myth of human sin and redemption ... But Darwinism short-changes us as regards our biological natures. We are every bit as co-operative as we

are competitive, as altruistic as we are selfish, as creative and playful as we are destructive and repetitive. And we are biologically grounded in relationships which operate at all the different levels of our beings … These are not romantic yearnings and utopian ideals. They arise from a rethinking of our biological natures that is emerging from the sciences of complexity.

(pp. xii–iv)

SORTING OUT THE LEVELS

When the idea of Gaia was first introduced, one of the things that shocked scientists about it was the way in which it clashed with this individualistic picture, which they were then accustomed to think of as particularly scientific. It seemed to them that they were being asked to accept an idea of organisms working cosily together to improve their environment, an idea which was incompatible with their evolving by cut-throat competition. Over-dramatising both these stories, critics asked whether Gaian thinking supposed these rival entities to form committees and plan climate change together?

This conflict does not really arise because the two processes take place at different levels. At the local level, organisms do indeed compete with one another and with neighbouring species. But one of the ways in which they compete is in finding ways of improving their environment, features which alter it – say, by making it warmer or wetter – in a way that helps them to survive. As Lovelock says:

If, in the real world, the activity of an organism changes its material environment to a more favourable state, and as a consequence it leaves more progeny, then both the species and the change will increase until a new stable state is reached. On a local scale adaptation is a means by which organisms can come to terms with unfavourable environments, but on a planetary scale the coupling between life and its environment is so tight that the tautologous notion of 'adaptation' is squeezed from existence. The evolution of the rocks and the air and the evolution of the biota are not to be separated.

Our interpretation of Darwin's great vision is altered . . . It is no longer sufficient to say that 'organisms better adapted than others are more likely to leave offspring'. It is necessary to add that the growth of an organism affects its physical and chemical environment: the evolution of the species and the evolution of the rocks, therefore, are tightly coupled as a single, indivisible process.

(Lovelock, *The Ages of Gaia*, Oxford, Oxford University Press, 1988, pp. 33–4, 63)

Such improvements can help others without damaging those who make them, because they expand the living-opportunities available to all in the system. That is how life was able to spread so widely over the planet in the first place. It makes no difference to this result which of the competing species got ahead of another in a particular development because this is not a zero-sum game. An obvious example of such a situation is the tropical rain forest which continually absorbs and recirculates rain. As Tim Lenton says:

A trait that brings the resulting organism closer to the optimum growth conditions will spread. Such a trait is, by definition, 'Gaian'. In contrast, a mutation in an 'anti-Gaian' direction will have its spread restricted by putting the organism responsible at an evolutionary disadvantage . . . There are many examples of living plants altering climate to their own benefit. Ecosystem-level environmental feedbacks must be understandable in terms of natural selection . . . Ecosystems that have stabilising feedback will tend to persist and spread, whereas ecosystems that develop destabilising feedbacks will tend to collapse and disappear.

(Timothy M. Lenton, 'Gaia and Natural Selection', *Nature*, vol. 394, 30 July 1998, pp. 444–5)

LIVING?

Another feature which alarmed some scientists was the use of the word 'life'. Is it legitimate, if one accepts this way of thinking, to say that the planet itself is in some sense alive? Obviously this is a verbal question, but it raises very interesting considerations about the way in which this concept applies to entities operating on different scales.

One objection made to calling the earth alive was that nothing can be alive unless it reproduces, and of course planets do not go out and mate with other planets. The development of the biosphere has therefore not proceeded, like that of a particular species, by the mutation and selection of planetary genes.[1] But it is not obvious that reproduction of this kind has to be a necessary condition for an entity's being considered alive. Spermatozoa, for instance, are commonly thought of as alive, since they visibly swim around. They are unquestionably part of the process of life. But they do not mate with other spermatozoa to produce young and allow of natural selection between their progeny. Scientists can distinguish between living and dead spermatozoa without having to suppose them capable of reproducing *on their own scale*. Similarly, the distinction between Mars and Venus as dead planets and the earth as a living one can, as we have seen, be made by clear and relevant marks without any reference to planetary reproduction.

Perhaps we should think of life, like 'order', as something that can be present in different ways in units of different sizes. Enquiring about this, Lewis Thomas comments:

Item. I have been trying to think of the earth as a kind of organism, but it is no go. I cannot think of it in this way. It is too big, too complex, with too many working parts lacking visible connections. The other night, driving through a hilly, wooded part of Southern New England, I wondered about this. If not like an organism, what is it like, what is it *most* like? Then, satisfactorily for that moment, it came to me: it is *most* like a single cell.

(Lewis Thomas, *Lives Of A Cell*, London, Futura, 1976, p. 4)

– an analogy which he proceeds to develop. All this raises the question of what elements the notion of life actually involves, something that is really not simple. It is a complex concept which remains in many ways mysterious to us. Lovelock comments:

Take the concept of life. Everyone knows what it is but few if any can define it. It is not even listed in the [standard] *Dictionary of Biology*. If my scientific colleagues are unable even to agree on a definition of life, their objections to Gaia can hardly be rigorously scientific.

If we ask a group of scientists 'What is life?' they will answer from the restricted viewpoint of their own scientific disciplines. A physicist will say that life is a peculiar state of matter that reduces its internal entropy in a flux of free energy, and is characterised by an intricate capacity for self-organisation ... A neo-Darwinist biologist will define a living organism as one able to reproduce and to correct the errors of reproduction by natural selection among its progeny. To a biochemist, a living organism is one that takes in free energy as sunlight, or chemical potential energy, such as food and oxygen, and uses the energy to grow according to the instructions coded in its genes.

To a geophysiologist, a living organism is a bounded system open to a flux of matter and energy, which is able to keep its internal medium constant in composition and its physical state

intact in a changing environment: it is able to keep in homoeostasis . . . *Gaia would be a living organism under the physicist's or
the biochemist's definitions.*

(Lovelock, Gaia, *The Practical Science of Planetary Medicine*,
p. 29; emphasis mine)

The crucial point is that life is not an accident or an alien invader
but something which has grown out of the earth itself. The sharp
divisions we make across this continuum reflect academic
specialisations rather than unbreakable natural barriers.

> There is no clear distinction anywhere on the earth's surface
> between living and non-living matter. There is merely a hier-
> archy of intensity going from the 'material' environment of the
> rocks and the atmosphere to the living cells. But at great
> depths below the surface, the effects of life's presence fade.
> It may be that the core of our planet is unchanged by the
> presence of life, but it would be unwise to assume it.
>
> (Lovelock, *The Ages of Gaia*, p. 40)

He points out that the things which we think of as most clearly
alive often have parts which are not alive, just as the earth does.
Our own teeth, hair, nails and bones are largely dead, but all
these things are part of us, some of them necessary parts. Then
there are trees. The bulk of a tree – the heartwood – is not alive
and neither is the outside bark. There is just a thin layer of living
tissue under the bark and in the leaves. But these are all parts of
the living tree. Again, coral polyps are the only live part of a coral
reef, but they have built the reef and they form a whole with it.
A termites' nest, similarly, has been formed by its inhabitants
and is constantly being changed by their activity. Polyps and
termites cannot possibly be understood in abstraction from their
co-evolved homes. Neither can we.

PRIORITY QUESTIONS

Most of this discussion has dealt with the imaginative role of Gaian thinking, not with its implications for policy. But of course those issues are linked. A clearer, more realistic imaginative vision of the world is bound to make for a clearer sense of priorities. The lurid competitive myths which have recently coloured our views both on human social life and on evolution can obscure our real dangers completely. As I write, the main items in the news are typical ones: not (of course) how to use less fossil fuel so as to save the rain forests, but how to force the French to eat British beef which they believe to be tainted with mad cow disease and one more dispute in Northern Ireland.

Of course human beings naturally think like this much of the time. Local feuds fascinate us. But we don't have to be wholly imprisoned by them. Common dangers can shake us out of this narrowness. They are, it seems, beginning to do so. And when they do, it is very important that we should have other, non-competitive ideas available to give us a different perspective.

For about a century and a half, competitive ideologies have reigned more or less unopposed in our culture and the notion of the physical world as an infinitely exploitable oyster has been widely accepted. Social atomism and social Darwinism have been the romantic myths of the early capitalist age. They are the background assumptions which we now need to correct. What Gaian thinking can do is to help us to do this by seeing what is before our eyes rather than looking at these videos. It brings us up with a quite new force against facts that we have been told about already but have never really taken in.

Does it also change our view of what those facts are? In detail it probably will do this. Scientists are now using Gaian ideas (not always under that name) to investigate how the earth's maintenance systems work, and no doubt they will reveal new factors. But the main picture is before us already. It centres on global warming. The earth is suffering from a dangerous fever while the most powerful people on it keep piling on more and more blankets of greenhouse gas – the very stuff that life has had so much trouble keeping under control for so many centuries – thus making the fever worse. And there are so many of us humans now – so many more than the ailing earth can easily carry – that we can no longer rely on slight palliative measures. It is foolish to keep on 'buying time' and not using it.

What, in this situation, needs to be done first? This is a question about priorities. And the key to it is perhaps clearest in the image that I used earlier of an ocean liner which is beginning to sink – only (as we explain) not at our end . . . Of course it is understandable that we do not see the planetary danger. Other, more immediate evils constantly demand our attention. Conditions on the terrestrial ship are bad in a thousand ways and endless things need to be done about them. But if the ship sinks, curing those evils will not be much help. The message is not that we should value the health of the earth above human needs. It is that these are not alternatives. Without a healthy earth, humans cannot survive anyway. As Lovelock puts it:

> I find it bizarre in these circumstances that our environmental concerns are nearly all human and personal. We worry far more about some remote danger of harm from pesticide or unusual genes in food than we do about the grim inevitability of global warming and all the harm that it will bring. We drive heedlessly to the supermarket in our polluting cars where we buy organic, pesticide-free food for ourselves and our family. Our priorities are all wrong. Our demons of nuclear radiation and carcinogens from chemical industry are there but tiny and feeble compared with the monsters that endanger the earth and that we made.
>
> We should fear the effects of removing natural habitats with their ability to serve as global and local regulators. We should

stop all further habitat replacement by farmland. We might even need to encourage intensive agriculture if doing so saves land that can be set free to return to its natural state.

We should fear the consequences of changing the composition of the atmosphere. We need to replace as soon as possible fossil fuel energy production with solar, nuclear or any other large-scale non-polluting power sources.

(Lovelock, personal communication, December 1999)

These things come first. After them, he says, comes the need to preserve fresh water and to prevent the excessive accumulation of CFCs and similar substances that deplete stratospheric ozone and add to the thickness of the gaseous greenhouse.

All other environmental projects come after these. Of course that does not mean that they do not matter. They do matter, just as it would still matter to look after sick people on a sinking ship. But doing so is no substitute for plugging the leak. Of course Lovelock's relative tolerance of nuclear power does not mean that he ignores its dangers. It simply marks his sense of the far greater dangers – to humans as well as to everything else – that go with gross changes in the earth's climate and the overriding need to meet those dangers right away.

Is it actually possible for us to shift our priorities in this way? Does the new millennium, with its promise of change, perhaps make so drastic an alteration possible? Can it shake our deep and habitual short-termism? These curious lines across the calendar do help our thought, in spite of the nonsense that attends them. They serve to remind us that the world does move on, that our recent ways of living are not set in stone as eternal verities. Perhaps it is rather surprising that we need this reminder in an age where huge changes are already happening, an age where (to take trivial examples) businessmen in London now consult a Feng Shui expert about the site of their offices and are afraid to smoke in meetings, two things which they would have found unthinkable twenty years ago. It is not just an unreal piece of moralising to suggest that we should cut down the use of cars. In spite of the car-lobbies, the approaching prospect of gridlock is not actually popular and many European cities, such as Frankfurt, now live happily with streets that have been largely cleared of the nuisance.

Cars are indeed a most interesting and potent symbol of our changing concepts of freedom. In fantasy, television advertisements show cars as a way of liberating their (usually solitary) driver from all outside interference. A solitary car roams the landscape, achieving a bizarre kind of omnipotence. In the real world, however, it has unfortunately proved impossible to eliminate the other drivers so this dream-solipsism is disappointed. Everybody, therefore, tries to achieve their own private omnipotence, resulting in a lot of damaging stress, gridlock and road-rage.

Our recent method of handling of the planet has been rather similar and it is turning out no more successful. Perhaps it really is time for us to change it.

NOTES

INTRODUCTION

1   In an article called 'Why Memes' in *Alas, Poor Darwin!*, ed. Steven and Hilary Rose (London, Jonathan Cape, 2000) a book which also contains a full discussion of Evolutionary Psychology.
2   Richard Dawkins, *River Out Of Eden* (London, Weidenfeld and Nicolson, 1995) p. 19.
3   See *The Symbiotic Planet* (London, Phoenix Books, 1998).
4   From 'Thoughts for the millennium: Richard Dawkins' in *Microsoft Encarta Encyclopaedia 2000* (Microsoft Corporation, 1993–9).
5   Peter Strawson, *Freedom and Resentment* (London, Methuen, 1974), p. 1.

1 THE SOURCES OF THOUGHT (no notes)

2 KNOWLEDGE CONSIDERED AS WEED-KILLER

1   E.g. Book II, line 596–600 and lines 993–8.
2   In *Evolution As A Religion* (London, Methuen, 1985), opening pages and chapters 9–10.

3   On the close analogies between plumbing and philosophy, see my book *Utopias, Dolphins and Computers: Problems in Philosophic Plumbing* (London, Routledge, 1997), especially the opening chapters.

3 RATIONALITY AND RAINBOWS

1   P.B. Shelley, *The Defence of Poetry* (1820, published 1840). In *Political Tracts of Wordsworth, Coleridge and Shelley*, ed. R.J. White (Cambridge, Cambridge University Press, 1953), concluding sentence.
2   London, Penguin, 1998.
3   *Unweaving the Rainbow*, pp. 17 and 18.
4   B. Farrington, *The Philosophy of Francis Bacon* (Liverpool, Liverpool University Press, 1970), pp. 92, 96 and 62.
5   Bacon, *The New Atlantis*, Vol. 3, p. 156. Emphasis mine.
6   Descartes, *Principles of Philosophy* in Alquie (ed.) *Oeuvres Philosophiques*, p. 502, note.
7   Brian Easlea, *Witchcraft, Magic and the New Philosophy* (Brighton, Harvester Press, 1980), p. 182.

4 THE SHAPE OF DISILLUSION

1   John Ziman, *Reliable Knowledge: An Exploration of the Grounds for Belief in Science* (Cambridge, Cambridge University Press, 1978), p. 6.
2   John B. Watson, *Psychological Care of Infant and Child* (New York, W.W. Norton and Co., 1928), pp. 81 and 5–6.
3   H. Oldenburg, 'Publisher to the Reader' in Robert Boyle, *Experiments And Considerations Touching Colours* (London, 1664).
4   Stephen Prickett, *Coleridge And Wordsworth: The Poetry of Growth* (Cambridge, Cambridge University Press, 1970), p. 7.
5   *The Seasons*, 'Spring', 203–17.
6   Keats, Letter 64 to John Hamilton Reynolds, 3 May 1818.
7   Preface to the second edition of the *Lyrical Ballads, Wordsworth's Poetical Works*, ed. Thomas Hutchinson (Oxford, Oxford University Press, 1936), p. 935.

5 ATOMISTIC VISIONS: THE QUEST FOR PERMANENCE

1   London, Penguin Books, 1989.
2   In his *Physiological Epistles*, 1847. Quoted by John Passmore in *A Hundred Years of Philosophy* (Harmondsworth, Penguin, 1968), p. 36.

6 MEMES AND OTHER UNUSUAL LIFE-FORMS

1   London, Wildwood House, 1974, p. 67.
2   Oxford, Oxford University Press, 1976.
3   New York, Knopf, 1998.
4   *Consilience*, p. 136.
5   ibid., p. 50.
6   Richard Dawkins, *The Selfish Gene* (Oxford, Oxford University Press, 1976), pp. 206–7.

7 PUTTING OUR SELVES TOGETHER AGAIN

1   See for instance Jeffrey Gray's somewhat indignant assertion of a territorial claim for science against philosophy in the *Journal of Consciousness Studies*, vol. 2, issue 1, 1995, p. 8.
2   This is a main theme of his last book *Beyond Freedom and Dignity* (Harmondsworth, Penguin, 1973).
3   The systematic doubt by which he reached this conclusion is beautifully set out in brief in his *Discourse on Method*, parts 4 and 5 and more fully in his *Meditations*.

8 LIVING IN THE WORLD

1   See *Existentialism and Humanism* by Jean-Paul Sartre, trans. Philip Mairet (Methuen, London, 1948).
2   See the end section of *The Anthropic Cosmological Principle* by John D. Barrow and Frank R. Tipler (Oxford, Oxford University Press, 1986). I have discussed the world-view that lies behind these bizarre proposals in *Science As Salvation* (London, Routledge, 1992), pp. 19–29 and 195–218, and in *Utopias, Dolphins and Computers* (London, Routledge, 1996), Chapter 12.
3   My first book, *Beast and Man* (new edition Routledge, London, 1995), was largely concerned with the need to put ourselves together again by mending the Cartesian gap. It contains several discussions of Sartre's separatism (see under his name in the index). I have returned to the whole topic in *The Ethical Primate* (London, Routledge, 1995) and in an article called 'One World, But A Big One' in the *Journal of Consciousness Studies*, vol. 3, no. 5–6, 1996, pp. 500–15, which I have re-worked in Chapter 12 of this book.
4   See I. Eibl-Eibesfeldt, *Love And Hate* (Methuen, London, 1971), pp. 11–13 and 208–16.

5   David Hume, *Treatise of Human Nature*, book 1, part 4, section vii.
6   See for instance *Beyond Freedom and Dignity*, pp. 102–3 and 111–12.
7   *Principles of Human Knowledge* by George Berkeley, section 51, see also sections 37 and 38.
8   Hume, *Treatise on Human Nature*, book 1, part 4, section ii.
9   A point well brought out by Raymond Tallis in a sharp little book called *Psycho-Electronics* (London, Ferrington, 1994) and in *On The Edge Of Certainty* (London, Macmillan Press, 1999).
10 *The View from Nowhere* by Thomas Nagel (Oxford, Oxford University Press, 1986) p. 3.
11 Aristotle, *Nicomachean Ethics*, book 1, chapter 3, 1094b 1.24.
12 Lewis Carroll, *The Hunting of the Snark*, Fit the Second.

9 THE STRANGE PERSISTENCE OF FATALISM

1   E.J. Lowe, 'There are no Easy Problems of Consciousness' in the *Journal Of Consciousness Studies*, vol. 2, issue 3, 1995, pp. 266–72. Emphasis mine. Chalmers' article 'Facing Up To The Problems of Consciousness' appeared in the same issue of the *Journal*, pp. 200–20.

10 CHESSBOARDS AND PRESIDENTS OF THE IMMORTALS

1   Epictetus, *Discourses*, book 2, chapter ii: 'The Game Of Life'. Cf. Marcus Aurelius, *Meditations*, book 4, section 3; book 6, section 2.
2   Richard Dawkins, *The Selfish Gene* (Oxford, Oxford University Press, 1976) p. 215.
3   London, Simon and Schuster, 1994.
4   See Churchland, P. and P. 'Intertheoretic reduction: a neuroscientist's field guide' in John Cornwell (ed.) *Nature's Imagination* (Oxford, Oxford University Press, 1995), pp. 64, 71 and 74.
5   Vol. 2, no. 4, 1995.
6   Another intriguing case is the placebo effect in medicine, a puzzle well discussed by Patrick Wall in *The Science of Consciousness*, ed. Max Velmans (London, Routledge, 1996), pp. 162–81.
7   Harvard, Belknap Press, 1975, p. 3.

11 DOING SCIENCE ON PURPOSE

1   For a discussion of this vital but difficult concept, see G.E.M. Anscombe, *Intention* (Oxford, Basil Blackwell, 1957).

2   'Neuroscience and Folk Psychology: An Overview' in *Journal of Consciousness Studies* vol. 1, no. 2, Winter 1994, pp. 205–16.

3   On the melodramatic exaggeration of the genes' actual role in development, see Evelyn Fox-Keller, *Refiguring Life: Metaphors of Twentieth-Century Biology* (New York, Columbia University Press, 1995) especially the quotation on p. 27.

4   Tacitus, *Annals*, book XV, chapters 51 and 57.

5   In *The Poverty of Historicism* (1957) Karl Popper excluded all historical arguments on principle from the province of science. The difficulties which this ruling raised over the study of evolution were raised and discussed in a controversy in *New Scientist* in 1980 (vol. 87, pp. 215–17, 482, 511, 708–9 and 733–4). In his later book *Objective Knowledge: An Evolutionary Approach* (1972) Popper no longer imposed the ruling.

6   David Chalmers, 'Facing up to the Problem of Consciousness' in the *Journal of Consciousness Studies* vol. 2, no. 3, 1995, p. 210.

7   See *Causal Powers* by R. Harré and E.H. Madden (Oxford, Basil Blackwell 1975).

8   See his *The Rediscovery of the Mind* (Cambridge, Mass., MIT Press, 1992), chapter 7 and *The Construction of Social Reality* (London, Allen Lane, 1995) chapter 6.

12 ONE WORLD, BUT A BIG ONE

1   *The Construction of Social Reality* by John R. Searle (London, Allen Lane, 1995).

2   See my article 'The Game Game' (*Philosophy*, vol. 49, 1974) reprinted in my book *Heart and Mind* (London, Methuen, 1981).

3   'Facing up to the Problem of Consciousness' by David Chalmers, *Journal of Consciousness Studies*, vol. 2, no. 3, 1995, p. 210.

4   See *What is Life?* by Lynn Margulis and Dorion Sagan (London, Weidenfeld & Nicolson, 1995), an impressive book designed 'to put the life back into biology'. Its discussion starts boldly from where Erwin Schrödinger left the topic in his earlier book with the same title *What is Life?* (Cambridge, Cambridge University Press, 1967).

5   See his *Beyond Freedom and Dignity* (Harmondsworth, Penguin, 1973), p. 10.

13 A PLAGUE ON BOTH THEIR HOUSES

1 Karl Vogt, *Physiological Epistles*, 1847. Quoted by John Passmore in *A Hundred Years of Philosophy* (Harmondsworth, Penguin, 1968), p. 36.

2 See 'Thomas Henry Huxley: The War Between Science and Religion' by Sheridan Gilley and Ann Loades, *Journal of Religion*, vol. 61, July 1981, pp. 299–301. He viewed the conflict sceptically as a demonstration of the inadequacy of all theory. He considered however that, of the two positions, Humean idealism was actually the stronger. But he expressed his materialism so vigorously that his wider argument has been forgotten and he ranks (a century before Francis Crick) as the inventor of epiphenomenalism – a sad demonstration of the dangers of stating both sides of an argument.

3 'Rethinking Nature: A Hard Problem Within the Hard Problem' by Gregg Rosenberg, *Journal of Consciousness Studies*, vol. 3, no. 1, 1996, p. 76.

4 Thomas Nagel, *The View From Nowhere* (New York and Oxford, Oxford University Press, 1986) opening pages. I have discussed this passage in my book *The Ethical Primate* (London, Routledge, 1994), pp. 13 and 66.

5 See for instance 'Conscious Events As Orchestrated Space–Time Selections' by Stuart Hameroff and Roger Penrose in *Journal of Consciousness Studies*, vol. 3, no. 1, 1996, pp. 33–54.

6 I have discussed the ways in which this continuity can be understood without objectionable reductivism in *Beast and Man* (London, Routledge, revised edition 1995), and in more detail in *The Ethical Primate* (London, Routledge, 1994).

7 Lewis Wolpert, *The Unnatural Nature of Science* (London, Faber and Faber 1992), p. 121.

8 See *In The Beginning* by W.K.C. Guthrie (London, Methuen, 1957), chapter 3.

9 See R. Harré and E.H. Madden *Causal Powers: A Theory of Natural Necessity* (Oxford, Basil Blackwell, 1975).

10 'Consciousness – What Is The Problem and How Should It Be Addressed?', guest editorial by Jeffrey Gray, *Journal of Consciousness Studies*, vol. 2, issue 1, 1995, pp. 5 and 8.

11 See Werner Heisenberg, *Physics and Philosophy, The Revolution in Modern Science* (London, Penguin, 1990).

14 BEING SCIENTIFIC ABOUT OUR SELVES

1   See for instance Lewis Wolpert, *The Unnatural Nature Of Science* (London, Faber and Faber, 1992), chapter 7 on 'Non-science'.
2   See his *Rocks of Ages: Science And Religion In The Fulness Of Life* (New York, Ballantyne, 1999), pp. 133–71.

15 WIDENING RESPONSIBILITIES

1   There is now an enormous literature on environmental ethics, largely stemming from Aldo Leopold's book *A Sand County Almanack* (New York, Oxford University Press, 1949). Holmes Rolston III has defended the idea of the intrinsic value of nature in *Environmental Ethics: Duties to and Values in the Natural World* (Temple University Press, Philadelphia, 1988). J. Baird Callicott has surveyed the case for and against this position with special reference to the value of biodiversity in 'Conservation Values and Ethics' in *Principles of Conservation Biology*, ed. Gary K. Meffe and C. Ronald Carroll (Sinaur Associates Inc., Sunderland Mass., 1994). Callicott has also surveyed the views of various cultures on this matter in *Earth's Insights: A Multicultural Survey of Ecological Ethics* (University of California Press, Berkeley, California, 1994). I have myself discussed the clashes that have arisen between the different principles invoked in environmental ethics and in the defence of animals in 'Beasts Versus the Biosphere?', *Environmental Values*, vol. 1, no. 2, Summer 1992, pp. 113–22.
2   This is the well-known central doctrine of Kant's *The Groundwork of the Metaphysic of Morals*, translated by H.J. Paton under the title of *The Moral Law* (London, Hutchinson, 1948). Kant first establishes the status of rational beings as ends in themselves (p. 90) and then that of the community which they (ideally) form (p. 95). This is the kingdom of ends.
3   See *The Guardian* letters page for 26 March 1996. Emphasis mine.
4   It was Kant who spelt out this restriction with regard to animals. See his *Lectures on Ethics*, trans. Louis Infield (London, Methuen, 1930), p. 239. He was, however, ambivalent enough and worried enough about rejecting such duties to add the rather unconvincing rider that we must not ill-treat animals all the same, not because it was wrong but because doing so might make us ill-treat people. This piece of face-saving is the source of the Charity Commissioners' doctrine just mentioned and of other similarly implausible stories that are still

current. I have discussed it in my book *Animals and Why They Matter* (Athens, Georgia, University of Georgia Press, 1984), pp. 51–2.

5   It has been very well used in John Passmore's admirable book *Man's Responsibility for Nature* (Duckworth, London, 1974), which succeeds to a remarkable extent in avoiding pointless debates about words.

16 THE PROBLEM OF HUMBUG

1   See my book *Animals and Why They Matter* (Athens, Georgia, University of Georgia Press, 1984), chapter 5, especially pp. 62–4 about rights. On this point Stephen Clark usually follows a similar policy, though he is careful to explain the substantial points which lead people to insist on the word rights – as indeed I am too. See Clark, *The Moral Status of Animals* (Oxford, Oxford University Press, 1977), p. 34.

2   London, Routledge and Kegan Paul, 1983.

3   I have discussed these issues about the status of ideals in my book *Utopias, Dolphins and Computers: Problems in Philosophical Plumbing* (Routledge, London, 1996).

4   Bishop George Berkeley seems to have invented operationalism: see his book *The Principles of Human Knowledge*, sections 58 and 107–16, where he criticises Newton. The idea, developed in the nineteenth century by Mach, has played an important part in many recent scientific controversies. It is quite intelligible in the context of Berkeley's radical idealism, but it plays a rather odd part today in the Copenhagen interpretation of quantum mechanics, and in the suggestion – put forward by some sociologists of science – that science as a whole is just a 'social construction'.

17 INDIVIDUALISM AND THE CONCEPT OF GAIA

1   Plato, *Timaeus*, section 33.

2   Karl Popper, 'Of Clouds and Clocks' in *Objective Knowledge: An Evolutionary Approach* (Oxford, Oxford University Press, 1972), p. 222.

3   *Gaia, The Practical Science of Planetary Medicine*, p. 111.

4   For a review of recent discussions see Timothy M. Lenton, 'Gaia and Natural Selection' in *Nature*, vol. 394, 30 July 1998.

5   See for instance Keith Devlin in *Goodbye Descartes: The End of Logic and the Search for a New Cosmology of the Mind* (New York, John Wiley and Sons, 1997) and the entire works of Richard Rorty.

6   For examples, see John Passmore, *Man's Responsibility For Nature* (Duckworth, London, 1974), chapter 1.

7   This is why there is now a thriving *Journal Of Consciousness Studies* – something that would have been inconceivable twenty years ago.

18 GODS AND GODDESSES: THE ROLE OF WONDER

1   On the radical interdependence between their religious and scientific thinking see Margaret Wertheim, *Pythagoras' Trousers, God, Physics and the Gender Wars* (London, Fourth Estate, 1997), chapters 5 and 6.

2   This is the topic of his book *Gaia, The Practical Science of Planetary Medicine.*

3   *Gaia, The Practical Science of Planetary Medicine*, pp. 6, 11, 31.

4   *The Ages of Gaia*, op. cit., pp. 206 and 212.

5   By Paul Davies (New York, Simon and Schuster, 1984).

6   By Paul Davies (New York, Simon and Schuster, 1992).

7   By Leon Lederman and Dick Teresi (Boston, Houghton Miflin, 1993).

8   By Frank J. Tipler (New York, Doubleday, 1994).

9   *Pythagoras' Trousers*, op. cit., p. 100. For more about this amazing but highly influential sexual chauvinism see Brian Easlea, *Science and Sexual Oppression* (London, Weidenfeld and Nicolson, 1981).

10   Farrington, *Philosophy of Francis Bacon*, pp. 62, 92, 93; Spedding, *Works of Francis Bacon*, vol. 4, pp. 42, 373.

11   John Passmore laid out this problem admirably in *Man's Responsibility for Nature* (London, Duckworth, 1974) and it has continued to occupy environmental philosophers ever since.

12   John Rawls, *A Theory of Justice* (Cambridge, Mass., Harvard University Press, 1971), p. 512, cf. p. 17. I have discussed this remarkable move in *Animals and Why They Matter* (Athens, Georgia, University of Georgia Press, 1984), pp. 49–50.

19 WHY THERE IS SUCH A THING AS SOCIETY

1   I have discussed this fascinating point more fully in *Beast and Man* (London, Routledge Revised Paperback Edition, 1995), chapter 6.

20 PARADOXES OF SOCIOBIOLOGY AND SOCIAL DARWINISM

1   For Darwin's rejection of Spencer's approach, see his autobiography, *Autobiographies of Charles Darwin and Thomas Henry Huxley* (Oxford, Oxford University Press, 1974), p. 64.

2   See Richard Hofstadter, *Social Darwinism in American Thought* (New York, Braziller, 1959), p. 47.

3   Edward O. Wilson, *Sociobiology: The New Synthesis* (Cambridge, Mass., Harvard University Press, 1975). I have not multiplied quotations from Wilson and Dawkins here because I do not want to waste space on books that are so well known. But I have discussed them fully elsewhere, notably in *Evolution as a Religion* (London, Methuen, 1985), chapters 14 and 15, and in *Beast and Man* (London, Routledge, 1995), where the early chapters, especially chapter 4, deal largely (and sympathetically) with Wilson.

21 MYTHOLOGY, RHETORIC AND RELIGION

1   See Richard Dawkins, *The Extended Phenotype* (Oxford and San Francisco, W.H. Freeman and Co., 1982), pp. 234–7.

INDEX